THE PROPER CARE AND FEEDING OF ZOMBIES

Brains! Brains! Brains! Brains are tasty. Brains are squishy. I want brains. I need brains. B R A I N S ! Brains make me feel good. I am hungry for brains. I smell brains. Where's my foot? Brainsss. Brains are beautiful. I yearn for brains. Brains are warm. Somebody ate my arm. Is that a brain? I am hungry for brains again. Always hungry for brains. I enjoy eating deep fried brains. Brains make me feel warm and fuzzy inside. Brains are soft. I love brains. You love brains, we all love brains. Strange, I'm hungry again for more brains. Is that a brain pie? I smell brains again. Do you smell brains? Who has brains? Brains! Brains! Brains! Give me more Brains! I want more BRAINS! MUST HAVE MORE B R A I N S ! B R A I N S !

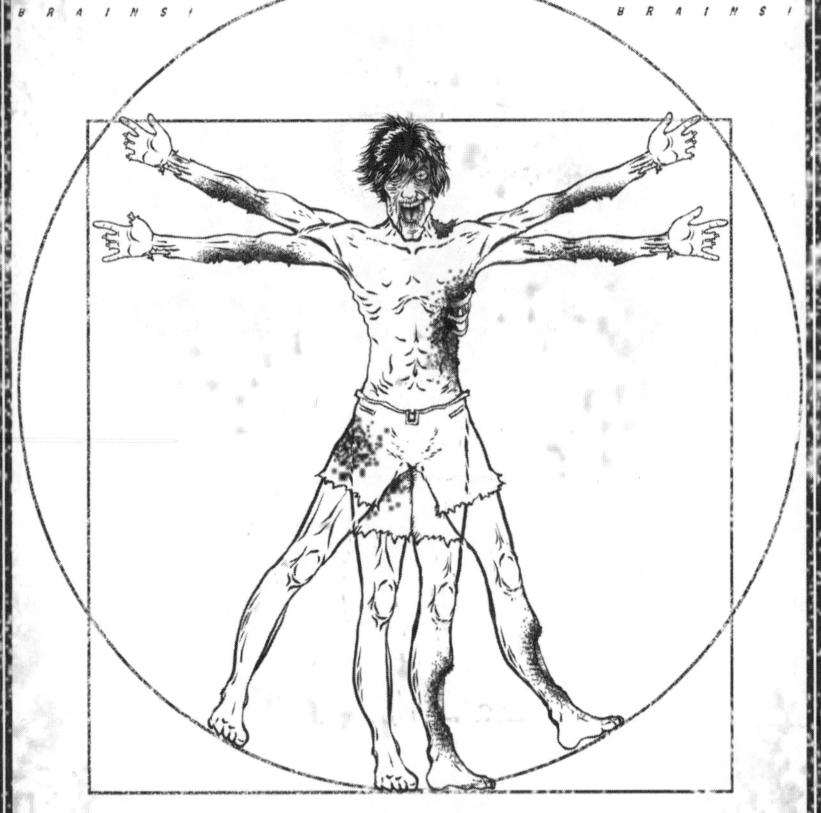

Brains! Brains! Brains! Brains are tasty. Brains are squishy. I want brains. I need brains. B R A I N S ! Brains make me feel good. I am hungry for brains. I smell brains. Where's my foot? Brainsss. Brains are beautiful. I yearn for brains. Brains are warm. Somebody ate my arm. Is that a brain? I am hungry for brains again. Always hungry for brains. I enjoy eating deep fried brains. Brains make me feel warm and fuzzy inside. Brains are soft. I love brains. You love brains, we all love brains. Strange, I'm hungry again for more brains. Is that a brain pie? I smell brains again. Do you smell brains? Who has brains? Brains! Brains! Brains! Give me more Brains! I want more BRAINS! MUST HAVE MORE B R A I N S ! B R A I N S !

THE PROPER CARE AND FEEDING OF ZOMBIES

■ ■ ■

A Completely Scientific Guide to the Lives of the Undead

MAC MONTANDON

WILEY

John Wiley & Sons, Inc.

This book is printed on acid-free paper. ♾

Published by John Wiley & Sons, Inc., Hoboken, New Jersey

Published simultaneously in Canada

Illustrations by Albert Lee, copyright © 2010 by John Wiley & Sons, Inc.

Design by Forty-five Degree Design LLC

Library of Congress Cataloging-in-Publication Data:

Montandon, Mac.
The proper care and feeding of zombies / Mac Montandon.
p. cm.
ISBN 978-0-470-64369-3 (paper : alk. paper); ISBN 978-0-470-90725-2 (ebk); ISBN 978-0-470-90726-9 (ebk); ISBN 978-0-470-90727-6 (ebk)
1. Zombies. I. Title.
GR581.M66 2010
398'.45—dc22

2010028337

Printed in the United States of America

10 9 8 7 6 5 4 3 2 1

■ ■ ■ ■ ■

For my dad, Henri, who showed me
how science can be fun.

CONTENTS

ACKNOWLEDGMENTS

I would like to thank the following people for their great help on this book. Jase Miles-Perez and James Weinheimer provided invaluable research assistance and too many wonderful ideas to mention. My dear old friends Jeremy Kasten and Aaron Ruby gave me excellent insights and an exhaustive list of must-watch movies. Matt Haber let me bend his ear about this for several months and is a fantastic source on just about anything—really, try him. The enthusiasm of my mom, my grandma, and my aunt Jamy; my cousins Zack, Eli, and Rose; and my friends Joanna, Aaron, April, Heather, Bill, Plato, Abe, Peter, Paul, Billy, Pinn, Katie, Mike, Tyler, and Dylan helped keep the project fresh and interesting. Professor Maurine Neiman, Dr. Mark Mahowald, and several other researchers and scientists were quite generous with their time, expertise, and PDFs. Connie Santisteban at Wiley is an extremely supportive and smart editor—thanks so much, Connie, for dreaming up this book and giving me the chance to write it. My agent, Anna Stein, is as relentless as the undead (in a good way) and responds to pretty much every e-mail with amazing, very much appreciated quickness.

For a book like this to exist, it needs a robust history of past masters: Thank you, Bela Lugosi, George Romero, Sam Raimi, and Max Brooks (and Mel, too!), among many others, for paving the blood-soaked way. Finally, as always, I truly couldn't and wouldn't have done it without the Big Three: my wife, Catherine, and my two daughters, Oona and Daphne. They are the best reasons ever to avoid being turned into a zombie.

INTRODUCTION

■ ■ ■ ■ ■

FANDOM starts young. I was not much older than seven when I first saw the original *Star Wars*, and I have wanted, at times desperately, to be Han Solo ever since. Zombie fandom can strike at an even younger age than that.

My daughter Oona was five and a half years old when she came home from school one day and excitedly told me about a new game that was all the rage on her kindergarten playground: Bye Bye Zombie! This was before I'd mentioned working on the book you are now holding, which means I hadn't influenced the test subject (my daughter). She arrived at her fandom all by herself—with the help of a few rambunctious little classmates capable of impressive and alarmingly real-looking zombie imitations—and her passion ran deep. She played it all the time throughout the year. It was also complex: in the game a player could choose to be a zombie, a human, or an animal; the zombies could eat the humans but not the animals, and the humans had guns with which to battle the undead. (Now that I think about it, I'm not sure what the animals did.)

Fortune cookie wisdom says that if you love someone, you should set them free. Most people ignore this advice. Instead, if they love someone, they stammer and sweat and eventually stalk their love interest until the person relents and finally, miserably agrees to go on a date. One. Date. The suffocating attention continues until the less than happy couple is married, knocked up, and in it for life.

In this way, a person in love is like the ultimate fan—in it for life. Zombie fans are arguably the most committed of all. Or, at any rate, I will put their collective passion up against that of fans of vampires, werewolves, mummies, and Oprah any day. When it comes to supernatural or metaphysical creations, few if any creatures inspire collective crushing like zombies. There's a good reason for the intensity of our feelings: for when we love someone, not only do we never want to set them free, we want to know every last possible thing there is to know about the person. Or the zombie. The ultimate zombie fan wants to know everything that can be known about our undead brethren—from their speed (fast or slow) to their diets (brains, brains, brains) to their sleep habits (hardly at all). The ultimate fan knows that a closer examination of how the other half lives (or not) is well worth a scientific explanation.

By now we know a good deal about our favorite ghouls. Thanks to the incisive efforts of some of our most gifted writers, filmmakers, and Italian math whizzes (Max Brooks, George Romero, and David Cassi, to name a few), we know what a zombie apocalypse looks like, how radiation from a fallen space satellite can cause an outbreak, and what modeling can tell us about escaping the brain-licking beasts. (Speaking of, if you're hiding in the basement right this minute and need to kill some time while waiting out the zombie horde, you may enjoy the "Killing Time" bits sprinkled throughout the book. They are, as the name suggests, really good for killing time, if not zombies.)

But surely we don't yet know everything there is to know. After reading this book, you still won't know everything there is to know about zombies. As my daughter's game underscores, zombies are complicated folk. But hopefully you will have learned a lot and will have had fun along the way.

So, thanks, zombies! While it's true that your bite is not exactly a love bite, our love for you goes on nonetheless. We love your lumbering ways, your gruesome moans, your flaking faces, and your appetite for life. Darn it, zombies, we love practically everything about you! (That slurping of our spleen thing is not too cool—please do something about that, if you don't mind.) More than other monsters, zombies are close to being us. They are us, in fact, just a little more decomposed. And slobbery. So of course we cherish them, as we cherish our friends, neighbors, and ourselves. Unfortunately, however, we still have to shoot them in the head.

Zombie Quiz

▪ ▪ ▪

If you are unabashedly unmatched in your undead trivia knowledge, you will undoubtedly get a kick (to the sternum region) out of this quiz. For the rest of you, there's always cheating.

1. What was to blame in the original *Night of the Living Dead* for the dead living that night?
   A. Ancient Indian burial ground
   B. Ancient Pakistani burial ground
   C. Radiation from fallen spacecraft
   D. Vegas
2. In the 28 Days series (excluding the movie of the same title with Sandra Bullock), the easily communicable zombifying virus was:
   A. Rage
   B. Slayer
   C. Anthrax
   D. Megadeth
3. According to Caribbean folklore, what is a surefire way of returning a zombiefied individual to his/her/its senses?
   A. Grooving to smooth island jams
   B. Andrew McCarthy and Jonathan Silverman–related hijinks
   C. Eating salt
   D. A shot of strong rum

4. "I shall raise up the dead and they shall eat the living. I shall make the dead outnumber the living." This proclamation by the Babylonian goddess Ishtar is chilling (though not nearly as chilling as the 1987 comedy *Ishtar*). What was Ishtar the goddess of?
   A. Love, war, and sex
   B. Love and Warren Beatty
   C. Canoeing, Ned Beatty, and backwoods hillbilly sex
   D. Earth, wind, and fire

5. In Norse mythology the re-animated corpses of fallen warriors possessed superhuman strength and a penchant for human flesh. What were they called?
   A. Draug
   B. Draugfüd
   C. Draughaus
   D. Draugibag

6. Harvard ethnobotanist Wade Davis wrote which nonfiction book about Haitian zombie culture and voodoo, which was later turned into an awesome movie?
   A. *The Squid and the Whale*
   B. *The Serpent and the Rainbow*
   C. *The Falcon and the Snowman*
   D. *The Pit and the Pendulum*

7. *The Zombie Survival Guide* and *World War Z: An Oral History of the Zombie War* author Max Brooks is the son of which show business couple?
   A. James L. Brooks and Valerie Harper
   B. Albert Brooks and Julie Hagerty
   C. Mel Brooks and Anne Bancroft
   D. Brooks and Dunn

8. In medieval Europe there were numerous documented cases of "revenants"—the deceased bodies of the wicked or unrighteous coming back to haunt the living. In Jewish folklore the accursed spirit would return and inhabit the living. What these were called you could tell me, maybe?

A. Dybbuk
B. Golem
C. Sajak
D. Trebek

9. What parasite, bred in the intestines of cats, can take over a rat's brain and alter it in such a way as to make it more susceptible prey? (Disturbing hint: A form of this parasite can also be found in nearly half of the human population.)

A. *Toxoplasma gondii*
B. *Toxoplastique candy*
C. *Toxmohandas gandhi*
D. *Toxmoesha brandy*

10. Haitian dictator François "Papa Doc" Duvalier was a practitioner of voodoo, which he supposedly used to control his own private "zombie army" known as the ________.

A. Sarlac Truffaut
B. Tonton Macoute
C. Womprat Shabaz
D. Chewie Vandross

11. The 1993 zombie romantic comedy—or zom-rom-com, colloquially—*My Boyfriend's Back* was directed by which Christopher Guest movie regular?

A. Eugene Levy
B. Bob Balaban

C. Parker Posey

D. Harry Shearer

12. In the 2009 Norwegian film *Død Snø* (*Dead Snow* for the non-Vikings among us), the young protagonists are terrorized by a murderous throng of the undead. To make matters worse, these particular zombies also happen to be________.

A. Vampires

B. Trumpeters

C. Morlocks

D. Nazis

13. The character Ash in Sam Raimi's *The Evil Dead* must defend himself from corpses and other demon-possessed entities brought on by incantations from the Egyptian *Book of the Dead.* This book is known by all of the following titles except?

A. Nyturan Demontah

B. Necronomicon Ex-Mortis

C. Nautica Delmonte

D. Naturon Demonto

14. Edgar Wright, Simon Pegg, and Nick Frost all worked on *Shaun of the Dead.* Years earlier they worked together on which British television show?

A. *Black Books*

B. *Spaced*

C. *Big Train*

D. *Danger! 50000 Volts!*

15. Scientists have fused silicone chips to viruses to create microscopic cyborgs that are able to operate even after the host being has died—creating a sort of living death. What are these tiny machines called?

A. Microbots
B. Nanobots
C. Acrobots
D. Wombots

Answers: 1. C, 2. A, 3. C, 4. A, 5. A, 6. B, 7. C, 8. A, 9. A, 10. B, 11. B, 12. D, 13. C, 14. B, 15. B

How's Your Score?

0–5 correct: You may want to check your pulse.

6–10 correct: You very well could be undead.

11–15 correct: You would be able to raise your arms in triumph, if only they hadn't been gnawed off by a zombie, meaning . . . oh, shit! Run, everyone, run!

ONE

KNOW THY ENEMY

■ ■ ■

What does a zombie brain look like? The neurobiology of zombies.

■ ■ ■ ■ ■

BRAINS. If there is a part of the anatomy that is more famously linked to zombies than brains, I'd like to know about it. As far back as we can remember—that is to say, 1985—the undead have hungered madly for the slithery matter found between the ears. In early big-screen zombie portrayals, though, the poor suckers appeared satiated as long as they could bite off their Shylockian pound of flesh from any old part of their victims' bodies. In 1985's *The Return of the Living Dead* (no relation to Romero's masterpieces), however, the monsters made it very clear that what they truly coveted for supper were brains. And lots of 'em. You may recall that it was in *ROTLD* that a doomed character named Tina bolted to a funeral-home attic to escape the zombific onslaught. Soon enough, however, Tina's boyfriend, Freddy, discovered her hideout. (Okay, he was really her ex-boyfriend at that point due to certain irreconcilable differences like, for example, the fact that Freddy was intent on slurping down Tina's neocortex and she wasn't yet ready to take the relationship to that level.) Upon discovering the hideout, Freddy commenced crashing through the ceiling to devour her. "I love you, Tina," zom Freddy half pleads, half demands, "and that's why you need to let me *eat your brains*."

Clearly zombies have a thing for a piping hot hypothalamus served just so with a side of corpus callosum. In fact, it's more than a thing; it's a gob-smacking obsession that has given rise to an enduring cultural meme with serious (atrophied, skin-flaking, rigidly unbendy) legs. In chapter 2 I will explore the effects of this resolutely undiverse diet on zombies and attempt to answer definitively the question that's haunted us these last twenty-five years: how much brains is too much brains, nutritionally speaking?

Before we discuss the zombies' amour for nibbling brains, I would like to turn the examining tables and discuss the zombie brain itself. To egregiously paraphrase Robert Frost: I'd now like to explore the brain less hungered for.

As you might suspect, there is not a surfeit of literature pertaining to the untangling of zombie brain wires. Happily, though, trailblazing work was done in 2009 by Dr. Steven C. Schlozman, an assistant professor of psychiatry at Harvard Medical School, a lecturer at Harvard School of Education, and—I doubt he'd take offense at my saying so—a Nerdus Emeritus at the School of Geeks.

Dr. Schlozman documented his findings in a faux-academic paper he titled "Ataxic Neurodegenerative Satiety Deficiency Syndrome" (ANSDS), which was drafted, tongue firmly in cheek, by five authors, three of whom were listed as "deceased" and another as "infected."

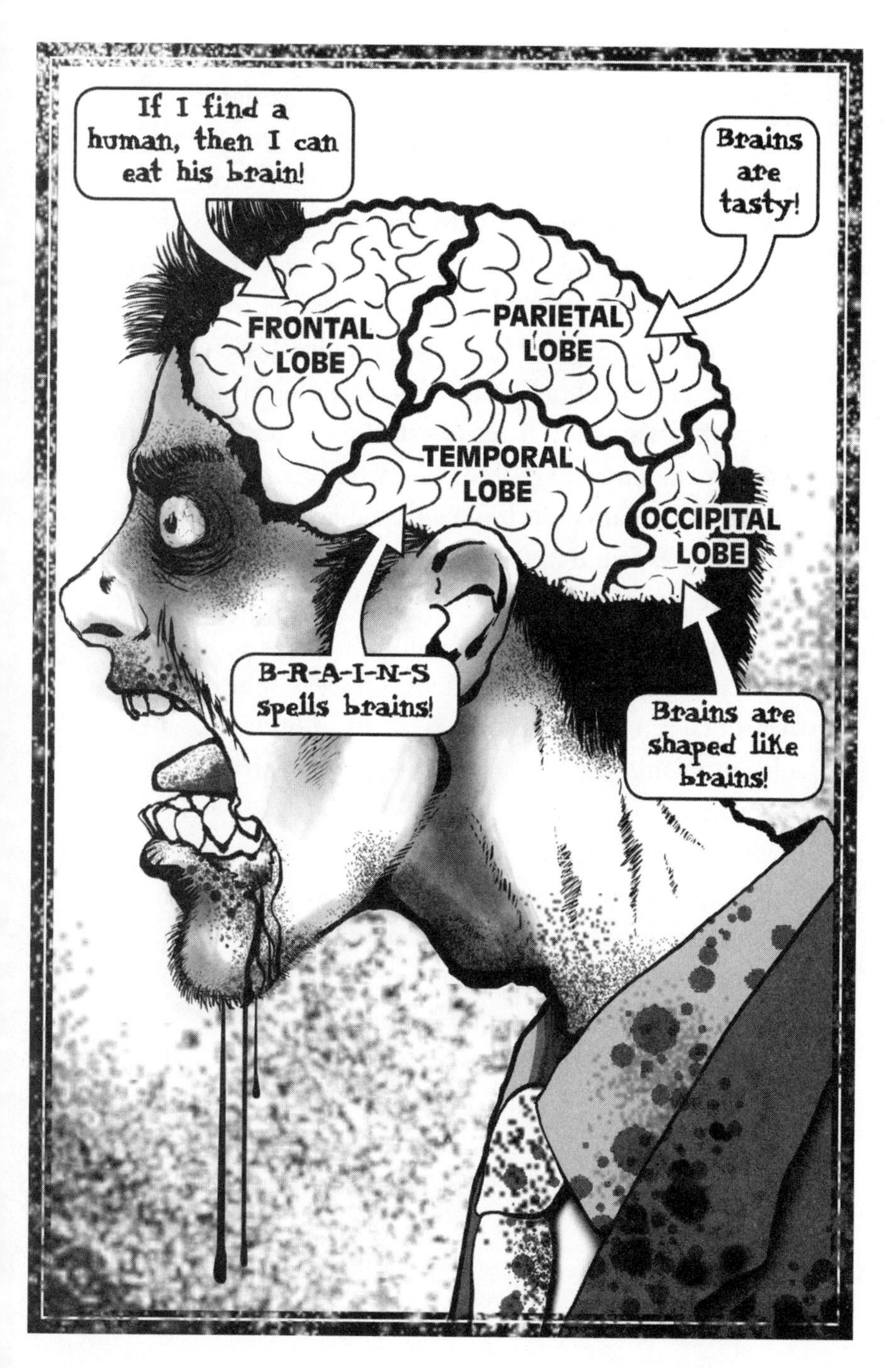
If I find a human, then I can eat his brain!
Brains are tasty!
FRONTAL LOBE
PARIETAL LOBE
TEMPORAL LOBE
OCCIPITAL LOBE
B-R-A-I-N-S spells brains!
Brains are shaped like brains!

After publishing his article, Dr. Schlozman presented his findings at a talk given as part of the Science on Screen lecture series hosted by the Coolidge Corner Theatre in Boston. The lab star's discoveries were startling, entertaining, instructive, and, crucially, lucid. Herewith, the major points of Dr. Schlozman's paper. The brain's frontal lobe is where something called "executive functioning" takes place. It's the part of the brain that helps us with abstract thinking and problem solving. It resides in the cerebrum, or cortex, which is the largest swath of brain real estate we have, the Russia of the brain if the brain were a map of the world. There are four lobes found in the cerebrum: the frontal lobe, parietal lobe, occipital lobe—all of which help us see and identify physical objects and, like, not bump into them; they give us our bearings—and temporal lobe. Of these lobes, I dare say the frontal is my favorite lobe. Why? In addition to helping us solve problems, the frontal lobe plays a big part in our ability to speak and to have emotions, which are both really great if you are a human who likes to communicate and feel stuff. The temporal lobe, by comparison, is quite good at maintaining our memories. And if you are anything like me, there are plenty of memories you would rather not keep at the ready, thank you very much, temporal lobe!

Dr. Schlozman suggests that zombies are not blessed with overly active frontal lobes. As anyone who has ever seen *Night of the Living Dead* can tell you, zombies essentially have only one approach to problem solving: Hungry? Eat! The frontal lobe also helps temper and control impulsivity. In his lecture Dr. Schlozman said that "clinically speaking, impulsivity is when you do something and if you had two more seconds you might not have done it." Zombies are nothing if not impulsive. Dr. Schlozman gives the following example of what can happen to humans with underdeveloped frontal lobes: some a-hole driver cuts you off on the freeway and you immediately flip him the bird. With just a dollop more frontal lobe activity, you might have realized that was not the best course of action—particularly if the other driver is a zombie or is wielding a sawed-off shotgun because he is hunting zombies.

Zombies can certainly see us. Case in point is the Hare Krishna flesh eater in *Dawn of the Dead* who stares longingly at the film's female star, a thin glass door and significant frontal lobe activity the only things separating them. Dr. Schlozman believes that zombies possess an adequately operating thalamus, the part of the frontal lobe that allows for sensory perception.

While executive functioning in zombies is severely lacking, the amygdala is robust, to say the least. *Amygdala* is Latin for "almond," and this part of the brain is indeed almond-shaped. The amygdala is in the lower, cruder part of the brain and is what drives our simple, unsophisticated emotions: rage, fear, and aggression.

Dr. Schlozman notes that another animal with a runaway amygdala is the crocodile, and he concludes that "you can't really be mad at zombies because that's like being mad at a crocodile. There's not much there other than amygdala—the essence of zombies is amygdala."

Put in those terms it's almost as though zombies can't help but feast on our flesh, so I'm with the good doctor here on pretty much everything, except that whole bit about being mad at them. If a zombie snacks on my wife or kids, I'm sorry but I'm going to be a little miffed.

Perhaps the biggest difference between humans and zombies—brainily speaking—is that human brains operate like a government is supposed to: through a series of effective checks and balances. If the amygdala is provoked and excited, there's the trusty anterior cingulate cortex to temper the rage, delay the impulsivity, and allow the frontal lobe to do some sober problem solving. Zombies, though, have a deficient anterior cingulate cortex and a weak frontal lobe—thus they routinely want to smash through walls to gnaw on our kneecaps. "Frontal lobes and amygdala are always talking to each other," Dr. Schlozman observes. "It's that balance between higher brain and lower brain that makes us humans."

Zombies and balance, meanwhile? Eh, not so swift there. The brain informs our ability to walk, run, jump, and move in a fluid manner. The regions of the brain responsible for these types of movements are the cerebellums (balance) and the basal ganglia (coordination). "If you've ever stayed out late drinking, you've experienced a lack of cerebellum," Dr. Schlozman says. In pretty much every zombie flick from 1932's *White Zombie* up until 2002's *28 Days Later* and beyond, the undead walk stiffly. Zombies are creaky, and that's because their cerebellum and basal ganglia are whatever the opposite of awesome is. They suffer from ataxia, which is the condition associated with cerebellar degeneration. According to the National Institute of Neurological Disorders and Stroke, "Ataxia often occurs when parts of the nervous system that control movement are damaged. People with ataxia experience a failure of muscle control in their arms and legs, resulting in a lack of balance and coordination or a disturbance of gait." Hello, zombies!

Remember earlier when I mentioned that Hare Krishna zombie in *Dawn of the Dead* and how he looks at the erstwhile news producer Francine (played by the consummate eighties actress Gaylen Ross) in a way that suggests a desire for contact more intimate than the mere teeth-on-forearm variety? He looks at her like he wants their souls to talk. Dr. Schlozman maintains that has to do with a very au courant neuroscience discovery called mirror neuron theory. Essentially mirror neuron theory indicates that the same part of the brain that reacts when, say, one takes a bite of pudding (or bicep) also reacts to seeing someone else take a bite of pudding (or bicep). What this means, Schlozman says, is that "we are wired to connect." That Hare Krishna, buried deep within his freaky zombie DNA, still has a delicate sense of connectedness to another human.

This might also help explain why zombies so often travel in packs. Those mirror neurons may be flickering more than firing, but they whisper nonetheless in the zombie brain. So when one zombie sees another zombie enjoying a plateful of human hands, he has brain activity equivalent to the one doing the actual eating and says to himself that he might also enjoy a round of digits. In that way zombies rely on one another to focus what might otherwise be horribly unfocused brains.

Ultimately, however, it's safe to say that zombies have a limited desire to connect with us or to understand our experiences. If that were not the case, wouldn't it be more difficult for them to engorge themselves on our parts? As further proof of this, Schlozman points to a passage in Max Brooks's definitive zombie novel, *World War Z*. It is there, during the bloody Battle of Yonkers, that we see the limitations of a zombie's mirror neuron activity. During an interview, a soldier who had fought in the battle recounts how wave after unstoppable wave of undead marched on him and his fellow soldiers. So the army blasted back with dead-eyed aggression—employing the most devastating weaponry in their arsenal—hoping if not to kill every last zombie then to at least give them pause, to cause fear. But that's like asking a nymphomaniac to lose desire—it just doesn't happen. And it didn't—the zombies were decidedly unmoved by the onslaught. They kept marching, stalking, crawling to fight.

"What if the enemy can't be shocked and awed?" Brooks asks in the book. "Not just won't, but biologically *can't*? That's what happened that day outside New York City, that's the failure that almost cost us the whole damn war. The fact that we couldn't shock and awe Zack boomeranged right back in our faces and actually allowed Zack to shock and awe us! They're not afraid! No matter what we do, no matter how many we kill, they will never, ever be afraid!"

So much for mirror neurons. The average human who sees his mates splattered then lying facedown in the mud is sure as hell going to think twice before advancing. Not so with zombies.

The last area of the brain discussed in Dr. Schlozman's lecture concerned the ventromedial hypothalamus, the tasty chunk of matter that tells us when we have had enough to eat. You can probably see where this is going. Zombies never stop eating—ipso facto, they likely have lesions on their ventromedial hypothalamus. Some studies have linked such a condition with chronic overeating and even obesity. But that research is decades old, and there is great skepticism about its validity. Let's assume for a second, because it is fun to assume, that there is a link: Why is it that zombies eat and eat and eat and never gain weight? Is it possible that all zombies are zombie models and possess the metabolisms of eight-year-olds? Probably not. I think it's more likely that zombies are, essentially, poster people for an Atkins-like, all-protein diet. No yucky carbs mucking up the works here! That and their commitment to exercise (twenty minutes a day? Try walking *every second of every day*!) is enough to keep the extra pounds off.

As a means of demonstrating what a zombie brain looks like in the real world, Dr. Schlozman invoked a medical phenomenon known well within the broader scientific community: the curious case of Phineas Gage. Gage was a twenty-five-year-old construction foreman working for the Rutland and Burlington Railroad company of New England when, on September 13, 1848, his life took a turn for the worse. According to a 1994 *Science* magazine story, that day Gage was overseeing a project that involved laying new rail tracks down in southeast Vermont, near the town of Cavendish. Gage's main job was to activate explosives in order to clear a path for the tracks. He had to drill openings in the rocks, fill the holes with detonating powder, cover that bit with dirt and sand, and then set the whole mess off by using a fuse and a tamping iron. Only, that day Gage was perhaps not paying as much attention to what he was doing as he should have been, and he began tamping the explosive material before any dirt or sand had been thrown on top of it. Well, you can probably guess what happened next—kaboom! The tamping iron, 3 centimeters in circumference and 109 centimeters long, blasted into his brain like the biggest bullet ever shot. Actually, the iron rocketed through his face and out the top of his skull. Somehow Gage lived through this and was soon able to regain consciousness. He was, miraculously, still in possession of his motor skills, able to talk with colleagues, and walk, albeit gingerly. Gage had survived.

"But he survived a different man," *Science* reported, "and therein lies the greater significance of this case. Gage had been a responsible, intelligent, and socially well-adapted individual, a favorite with peers and elders. The signs of a profound change in personality were already evident during convalescence under the care of his physician." The article went on to discuss how "in some respects, Gage was fully recovered. He remained as able-bodied and appeared to be as intelligent as before the accident; he had no impairment of movement or speech . . . and neither memory nor intelligence in the conventional sense had been affected."

From that day forward, Phineas Gage looked like us, but he was not us. The story soon took a dark turn: "On the other hand, he had become irreverent and capricious. His respect for the social conventions by which he once abided had vanished. His abundant profanity offended those around him. Perhaps most troubling, he had taken leave of his sense of responsibility."

Gage had experienced severe frontal lobe damage in the accident, not long after which he lost his will to connect. He no longer cared about his fellow human beings. He was, for all intents and purposes, a zombie.

Gage's doctor noted that "the equilibrium or balance, so to speak, between his intellectual faculty and animal propensities" was kaput. This brings to mind what a newscaster says about the rising, rib-chomping menace of *Night of the Living Dead*: "They look like people and act like animals." Put in even simpler terms, people who knew Gage said, "Gage was no longer Gage." The same might be said of you and me if we had lesions damaging our ventral and medial areas of our left prefrontal cortex.

The one huge difference between Phineas Gage and your average zombie is that Gage did not exhibit a significant uptick in the amount of human flesh he craved for supper. Post-trauma, he was not gurgling with rage. Why was that?

The authors of the 2004 scientific paper "The Neurobiology of Aggression," Pamela Blake and Jordan Grafman, think they might know the answer. Working out of Georgetown University's Department of Neurology, the two researchers suggest that frontal lobe lesions alone do not guarantee that the individual suffering the blow will become violent. "It may be that pathological aggression results from an interplay of impairment in areas of the brain responsible for modulating both emotional reactivity and also the response to environmental stimuli," Blake and Grafman wrote. "Neurotransmitter deficits may also contribute to the neurobiological basis for aggression. For example, abnormally low levels of serotonin have been found in the cerebrospinal fluid of violent offenders."

In other words—it's complicated. But what the authors do know for sure is that hyperaggressive humans rarely act that way for the first time in adulthood; instead, they have a history of violence. "With normal growth and development of the human brain," Blake and Grafman suggest, "the ability to suppress aggressive behaviors increases while our impulsive aggressive tendencies diminish, and the terrible twos evolve into the age of reason." Obviously, they've never seen *Dead Alive.*

But that does raise an interesting point, which is so long as one's brain develops normally, lesion free, and with all the proper checks and balances in place, there is no reason to think that one will suddenly turn into a zombie. Unless, of course, one is snacked on by the undead, acquires a mysterious virus, or encounters unsavory radiation (more on that later). As we noted earlier in this chapter, in many ways it all comes down to the parts of the brain that make us pause just a moment before flipping off a fellow driver, no matter how rudely they've swerved in front of us.

With that in mind, let's return for a moment to the area of the brain that plays the biggest part in the neurobiology of zombies: that slippery sliver of membrane, the amygdala—the almond. For while its relation to zombie neurobiology is simple on the one hand—as we've seen, an unchecked amygdala is quite a dangerous thing—some studies muddy the waters a bit. One study in particular could change the way we think about the brains of the undead. In a 2004 issue of *Psychological Science* magazine, researchers from Harvard, Yale, and the University of Toronto collectively published a report suggesting that the amygdala plays a part in unconscious race bias. For the study, the researchers flashed a series of photos of faces before thirteen white subjects. The faces were exposed to the subjects subliminally, in superfast thirty-millisecond flashes. What the researchers found was that there was greater amygdala response when a black face was shown to the study's participants than when a white face was revealed. Interestingly, when the photos were shown for longer periods of time, other parts of the brain—namely areas of the frontal cortex—kicked in and soothed the inflamed amygdala.

So what does this all mean for zombies? If, as Dr. Schlozman suggests, "the essence of zombies is amygdala," then why are zombies not more racially motivated?

That's hard to say for sure. Certainly in the two films that endure as unquestionable classics of the genre, *Night of the Living Dead* and *Dawn of the Dead*, if the zombies have any racial bias in them, they completely fooled us. In both of those flicks, the de facto heroes are young black men, Ben and Roger, respectively. And in a more contemporary example, Will Smith is the last human standing in *I Am Legend*. (Purists will argue that the infected monsters of that film are hardly zombies, based primarily on the speed at which they get around the deserted, desolate city.) Yes, he's Will Smith and he gets about $40 million per film, so of course he isn't going to get killed off early. But, still, the zombies didn't appear to make an extra effort to maul him.

What is more likely is that zombies, rather than seeing the world in black and white, see it as them versus us. The undead versus the living. (In this scenario, zombies are totally hip and postracial—the perfect monsters for the Obama age, Glenn Beck be damned!) A report last year in the *Journal of Neuroscience* makes just such a claim. There researchers announced that the region of the brain associated with feelings of empathy—the anterior cingulate cortex—is more active when one observes a member of one's own social group in pain than it is if one witnesses a stranger hurting. Zombies, with their degenerating anterior cingulate cortexes, clearly aren't capable of much empathy to begin with, but the little bit they can muster goes toward other zombies. That's who is in their social group, after all. They could give a damn what happens to humans—as the Evil Dead franchise makes perfectly plain.

It is, I submit, not a coincidence that literature from some of history's greatest thinkers brushes up very close to Dr. Schlozman's assessment of the zombie brain. Perhaps no book comes with as strong a whiff of zombific analysis as Descartes' *Discourse on Method*. In thinking about what makes a person a person the seventeenth-century philosopher famously arrived at this extremely neat tautology, "I think, therefore I am." The undead inversion of this notion would be something more along the lines of, "I don't think, therefore I must sauté your spleen."

A closer examination of this Cartesian principle underscores the way in which it can be argued that the brain and it's intimate, metaphysical cousin, the mind, are really all that separate us from them. Check out the following passage from *Discourse*, while substituting the word *zombies* for *machines*.

> If there were machines bearing the image of our bodies, and capable of imitating our actions as far as is morally possible, there would still remain two most certain tests whereby to know that they were not

> really men. Of these the first is that they could never use words or other signs arranged in such a manner as is competent to us in order to declare our thoughts to others.

Machines and zombies cannot communicate with words—check. And the second test?

> The second test is, that although such machines might execute many things with equal or perhaps greater perfection than any of us, they would, without doubt, fail in certain others from which it could be discovered that they did not act from knowledge, but solely from the disposition of their organs: for while reason is an universal instrument that is alike available on every occasion, these organs, on the contrary need a particular arrangement for each particular action; whence it must be morally impossible that there should exist in any machine a diversity of organs sufficient to enable it to act in all the occurrences of life, in the way in which our reason enables us to act.

Machines and zombies are not capable of the same level of moral complexity as we humans are because of their organs and stuff—check. From this, Descartes deduces, "Again, by means of these two tests we may likewise know the difference between men and brutes."

And so, in short, zombies may look like us and act like us and even, at times, move like us. But they are not us because their minds—their brains—not only don't work like ours, but they aren't capable of working like ours. Shock and awe, indeed!

In the magnificently twisted mideighties gore-fest *Re-Animator*, the dean of a spooky university has come back to life, thanks to an injection of a secret serum derived by a ghoulish student named Hubert West, played by the uniquely skeevy actor Jeffrey Combs. The dean, upon reawakening (he himself is the victim of a thrashing at the hands of a re-animated cadaver) is but a grim, unblinking, blood-spitting shell of a man. A zombie in a business suit.

The school's star professor and surgeon, Dr. Carl Hill, takes possession of the zombie dean and locks him in a padded cell, ostensibly to observe and attempt to cure him, but really to turn him into an undead slave. The dean's daughter, a fetching, willowy young coed with a blond bob, is, naturally, desperate for someone to cure her dad. She goes to Dr. Hill to plead for his help. While discussing the situation in his darkened office, Dr. Hill begins to tell the daughter how he intends to help the dean. "I want to take a look at the right frontal lobe," Dr. Hill says.

Dr. Hill, we later learn, is the creepiest cat of them all—really just an all-star sicko. But as we now know, his neurobiological instincts are impeccable. He absolutely should take a look at the dean's right frontal lobe, and stat!

For it is there, imbedded in the mysterious tissue of the brain, that humans find their humanity. Without a vigorously engaged frontal lobe, the amygdala can run wild. And the essence of the zombie brain is, as Dr. Schlozman noted, amygdala.

Now that we have a sense of how the undead brain works, let's take a peek at an entirely different aspect of the brain—how it tastes (how's that for a transition?). That is, historically, the dish zombies have fancied most. And they can seemingly eat no end of brain. Or foot, forearm, or calf. Makes one wonder: how much brain is too much brain? Can zombies live on flesh alone?

Those are the questions we will take up next. Dig in!

TWO

SERVE WITH A CHILLED PINOT GROSS

■ ■ ■

The benefits and hazards of an all-brain-and-human-flesh diet.

■ ■ ■ ■ ■

THE very first line of *Pride and Prejudice and Zombies*, Seth Grahame-Smith's robust rewriting of the classic Jane Austen novel, provides an excellent jumping-off point for this chapter: "It is a truth universally acknowledged that a zombie in possession of brains must be in want of more brains." Quite.

On a related note, there's a good chance you've heard this annoying adage more times than you'd like: You are what you eat. I detect at least a few obvious problems with that expression, but the biggest one is that it's not really true. Take today, for instance. So far I've eaten an apple, two delicious leftover roast pork tacos that my wife made for dinner last night, about forty-five cups of coffee, give or take, and a few glasses of water. Does that make me an apple-coffee-water-pork taco? Don't answer that.

Instead, answer this: Can you think of a truer example of that annoying adage than a zombie? No, you cannot. Zombies, after all, eat more or less only brains and human flesh. And zombies are brains and human flesh! Aha!

So while it's not really true for most of us that we are what we eat, it is actually true for zombies. Which is exactly why we should spend a little time here looking at what zombies are eating when they are eating human brains and flesh.

Now, unlike chapter 1, where we had a real-life scientist unpacking the neurobiology of the undead, we're going to be hard-pressed to find a legit academic squirrelly enough to have studied and then published findings about subsisting on people (and film school professors pontificating on *Soylent Green* don't count). In this context our improvisation will take us into some captivating and even dangerous realms of discovery: a culture of people whose diets are about 90 percent meat, studies of lesser-known brain diseases, and, naturally, cannibalism.

Let's start with cannibals—after all, given the choice they would totally start with us. Humankind has a long and complicated relationship with cannibalism. Most folks are aware of a few famous examples of hot person-on-person-on-plate action, like serial killer Jeffrey Dahmer who had an appetite for those he destroyed, or the famously unlucky westward travelers of the late 1840s, the Donner Party, who became lost in the snowy Sierra Nevada mountain range and had little choice for survival but to nibble on their neighbors.

It turns out, however, that we may have a much longer and more complicated cannibalistic history than even those gruesome examples suggest. (Maybe zombies are actually on to something.) In a 2003 study published in *Science*, Doctors Simon Mead and John Collinge discovered a genetic code

suggesting that people all over the world have been eating other people for a very long time. How long is hard to say, but Mead and Collinge based their research on a tribe called the Fore from eastern Papua New Guinea, who maintained a form of cannibalistic funeral practices from the late nineteenth century until a ban on such behavior was handed down by Australian officials in the mid-1950s.

For the Fore, there was nothing unsavory about what they were up to. In fact, in a paper written by the researcher Mike Alpers at Australia's Curtin University we learn that this "mortuary practice of consumption of the dead" was done in part to help "free the spirit of the dead."

Alpers's paper was focused on a rare brain disease called kuru, which, unfortunately, can be one of the nasty side effects of freeing the spirit of the dead. Kuru, as Alpers tells us, is "a fatal neurodegenerative disease with a subacute course lasting, on average, 12 months," and for which there is no cure. (As a side note, Kuru is also, truly, a former San Francisco rock group that once penned a tune called "Brain Bleeding Cannibal Core." So there's that.)

Like other neurodegenerative diseases—such as Creutzfeldt-Jakob disease and mad cow—kuru stems from a prion problem. A big one. Prions are proteins in the brain that mysteriously contort themselves into a mutant state, then infect other proteins, and then, in turn, the brain tissue. Thus kuru—which translates to the "laughing or trembling disease" for the effects it can have on the demented brain—is a class of infection known as transmissible spongiform encephalopathies. That is, it shreds brains like a sponge and can be transmitted to others.

The Fore transmitted kuru, it is believed, when tribespeople (particularly women and children) ate handfuls of the brains of the infected dead. Kuru is considered a progressive cerebellar disease, meaning the deterioration of the cerebellum occurs gradually. Early symptoms include headaches and joint pain, which then give way to full-blown cerebellar ataxia, which, as Dr. Schlozman noted in chapter 1, causes an infected person to move stiffly and unsmoothly, much like a *Day of the Dead* extra.

In studying the Fore, Doctors Mead and Collinge examined DNA samples from thirty older Fore women who had attended cannibalistic rituals. The samples showed that the majority of those women had a kuru-combating genetic signature that tribespeople who had not been cannibals did not possess. This is a little like building up antibodies to protect oneself from the common cold—only much more deadly.

Duly inspired, Mead and Collinge sought out samples from a global rainbow of peoples and discovered that every single ethnicity other than the Japanese were in possession of the same genetic signature that was found in the Fore women. The two then put forth a fascinating and controversial idea: they said that the DNA samples they'd analyzed indicated the presence of prion diseases throughout human history, and they believed that cannibals had passed along the disease. The disease-fighting genetic signature they had seen in the samples was proof of this, Mead and Collinge argued, writing that there existed "strong evidence for widespread cannibalistic practices in many prehistoric populations."

However, as the *New York Times* reported in covering Mead and Collinge's findings, they could not definitively say that it had been infected animals rather than infected people that had played a part in providing the genetic signature in question. Dr. Mead mostly dismissed this by telling the *Times* that "there is no animal prion disease that appears to cross the species barrier so easily and dramatically" as that of the human equivalent.

A thornier issue was raised by anthropologists such as Dr. William Arens at the State University of New York–Stony Brook. Dr. Arens suggested that cannibalism is not nearly as popular as Mead and Collinge would have us think and that it is often a specter raised in hopes of defaming a person or a group.

Wherever the truth actually lies, I think we can see how all of this impacts zombies. Undead brothers and sisters better be careful when they go around munching brains if they don't have that protective genetic signature in place. Of course, it could be argued that for zombies it is already too late—they are, it would seem, deep into some sort of prion disease or other. They certainly exhibit many of the physiological attributes of such diseases. In addition to the aforementioned cerebellar ataxia situation, kuru comes with tremors that are exacerbated at high-stress moments, and during the advanced stages of the disease, a loss of memory is experienced to the point that the infected may not recognize their own parents. It is then that a zombie cannot be morally in the wrong if he were to, say, devour his dad.

Perhaps the most threatening prion disease for zombies to be aware of is one grimly called deadly insomnia. As a 1999 BBC report described it, deadly insomnia "starts with protracted insomnia, leading to dementia, eventually reaching a state where sufferers cannot tell reality from dreams. They can expect to die between seven and 25 months of developing the condition."

So, clearly, zombies are at risk of contracting this particular prion disease because of both their lifestyles and their eating habits. Once awakened from, well, death, zombies never, ever hit the hay—they seem to have chronic and severe insomnia. And what with their taste for brains, zombies, like members of the Fore tribe, have a better-than-average chance of contracting deadly insomnia by feasting on toxic gray matter.

In some ways, I realize, it is not fair to warn zombies off of their all-brain-and-human-flesh diets with talk of prion diseases, the Fore, and kuru. For one thing, the funereal cannibalistic habits of the Fore only involved the brains of dead people. As long as zombies made sure they were munching strictly on fresh and healthy humans who themselves had eaten locally and organically, then the undead should be in the clear as far as brain-puncturing prion diseases are concerned. Additionally, there is no explicit evidence that I've seen that once zombies are infected—with zombiness—they are susceptible to other infections. The real dietetic concerns for a zombie have less to do with potential disease and more to do with the actual phenomena of a high-fat, high-protein, all-meat diet.

Which brings us to the Inuit of the Canadian Arctic. You might know them as Eskimos, but that's not exactly right. They are Inuit, and they reside on a shivery slice of land at the northeasternmost limits of North America—a place called Nunavut, in Canada, which translates to "our land" in Inuktitut. The truth is that the Inuit don't really eat only meat, but 90 percent of their diet is meat—and not just any meat. "Our meat was seal and walrus, marine mammals that live in cold water and lots of fat," Patricia Cochran told *Discover* magazine in 2004, recalling her experience growing up along the Bering Sea. "We used seal oil for cooking and as a dipping sauce for food. We had moose, caribou, and reindeer. We hunted ducks, geese, and little land birds like quail, called ptarmigan. We caught crab and lots of fish—salmon, whitefish, tomcod, pike, and char. . . . We ate frozen raw whitefish, sliced thin. The elders liked stinkfish, fish buried in seal bags or cans in the tundra and left to ferment. And fermented seal flipper, they liked that too."

In short, the Inuit like them some meat. Occasionally, in the summer, they also eat roots, greens, and berries. But mostly meat.

Unfortunately, for our purposes, practically the only meat the Inuit do not eat is pig. Some nutrition experts speculate that pig is closest to human flesh, nutritionally speaking. One highly regarded nutritionist (who asked that I not use his name because he didn't want to come off as some freak who thought about zombies all the time. Wait—what?) even told me, "Presumably, from a nutritional standpoint, there would be little difference between a human brain and a pig's brain." And so, given that the average human brain weighs about 3 pounds, or 1,300 to 1,400 grams, we can look at the nutritional value of 1,350 grams of pig brains to get a handle on what, exactly, zombies are eating. And what they are eating when they eat a single human brain is this: nearly 140 grams of protein, 125 grams of fat, lots of vitamin C, but barely any other vitamins and absolutely no carbs.

When that same nutritionist crunched the numbers he came up with some alarming stats for what the undead would be consuming were they to ingest a whole person. Hold on to your waistlines: a single serving of person—roughly a pound of flesh—brings with it about 4,900 grams of protein, 3,500 grams of fat, and a bit of vitamins A, B, D, and E. Also, lots of phosphorus and potassium, which would give zombies strong bones and account for why, when you bash them in the midsection with a shovel, they don't seem to mind that much. Finally, a pound of flesh contains a rather unhealthy 17,000 milligrams of sodium—ouch!

We'll return in a minute to look at the effects, both good and bad, of a diet (insanely!) rich with protein and fat. For now, let's stick with the Inuit to see how a largely all-meat diet works out for them.

For the most part, it works out really well. In fact, *Discover* noted that it is "surprising to learn how well the Eskimo did on a high-protein, high-fat diet." What they learned from studying the Inuit was that there are "no essential foods—only essential nutrients." Get those into your system and you can hunt down and gnaw up humans all day long without ever growing weary.

Anecdotally, in fact, it appears that the Inuit, largely due to their diet, are stronger and more vital than most average people. And they stay that way late into their lives.

In working on a 2002 paper called "Food and the Making of Modern Inuit Identities," written for the University of Alaska's anthropology department, Professor Edmund Searles, now at Bucknell, spent time with a family of four in Iqaluit, population about 6,000, which is the capital of Nunavut.

The oldest son of the family Professor Searles visited was a boy named Kelli. He spoke at length about the practical and nutritional benefits of the Inuit diet, as well as ways in which he believes it keeps his friends and family strong. Kelli told Searles that walrus meat is an excellent energy provider and stuffed as it is with vitamins and minerals, the food works as a sort of edible internal stove. "A pound of frozen walrus meat consumed before the start of a day of winter travel will keep you warm all day," Kelli said.

Kelli's brother, Ooleetoa, expanded on this notion and pointed to his dad, Aksujuliak, as the poster parent for the strengthening effect the Inuit eating habits can have on an older gentleman. Aksujuliak, you see, was sixty-eight years old at the time, and yet he was still the lead hunter of a local group. How had Aksujuliak maintained such a rigorous lifestyle? Searles wrote, "It was not unusual for him to consume several pounds of muscle and blubber in one sitting, or to drink several cups of fresh ringed seal blood. Ooleetoa attributed his father's vitality to his diet of Inuit food, including the muscle, blubber, and organs of ringed seal, bearded seal, walrus, Arctic char, and caribou."

This is already starting to sound like about half the cast of Peter Jackson's 1992 epic zombie splatter fest, *Dead Alive*, but it gets better (or worse, if you are unlucky enough to be on the business end of a zombie's chompers). Searles wrote, "Aksujuliak was known in Iqaluit for his ability to walk many miles without resting. It is for this reason he adopted the surname '*Pisukti*' ('land animal' or 'things that walk' in Inuktitut) in the 1960s."

Holy flesh-eating monsters! Land animal? Things that walk? These could just as easily be titles for future zombie flicks. Now, don't get me wrong, I am not suggesting that Aksujuliak is actually an undead ghoul who—were I to invite him into my home for a meal of pounds and pounds of whale blubber and pints of blood—would instead take a bite out of me. All I'm saying is that it is remarkable that in the upper, coldest corner of Canada there is a community of people who not only subsist on what is, more or less, a zombie diet, but they also exhibit behavioral traits that are sometimes associated with zombies—the superhuman strength, the unquenchable taste for flesh, and the ability to walk endlessly.

Additionally, like zombies, the Inuit avoid typical western problems with high-protein, high-fat diets—kidney stones and heart problems— by avoiding saturated fats, eating plenty of omega-3 fatty acids, and getting loads of exercise. As long as they are able to secure enough vitamins and minerals, they may have actually come up with one intensely healthy diet—and one zombies have wisely adopted.

Perhaps you've wondered how zombie children are so adept at picking up the eating habits and learning about the zombie diet from older zombies. Again, Searles has what sounds like a reasonably good answer for that when he described how Inuit parents pass on a love of the Inuit diet:

> [The] Inuit encourage their children to explore and express their autonomy and independence through eating. This is why many Inuit develop a passion for particular cuts of meat or for the organs of a particular animal, whether it is the shoulder meat of a ringed seal, the filet mignon of a caribou, the bone marrow of a caribou femur, or the heart of a ptarmigan.

(Or, Searles easily could have continued, the brains of a pitiful, unsuspecting, backlit schlub.)

As I suggested earlier, the only real concern for the Inuit—and zombies—vis-à-vis their diet has to do with consuming enough important nutrients. As it turns out, the Inuit are able to satisfy their nutritional needs in a variety of ways. Rather than securing vitamin A, as many of us do, from vegetables such as carrots and broccoli, those living in the far northeast receive theirs from the oils of cold-water fish and in the livers of other animals. Vitamin D is supplied to the natives from much the same sources. As for vitamin C, it is not quite as readily apparent how or where the Inuit are finding this valuable nutrient. And that's an important one to have, of course, as it is really good at warding off scurvy. Symptoms of that old-time disease include aching joints, blood vessel oozing problems, and general physiological and neurological decrepitude related to disintegration of connective tissue. All of which is not so helpful when you are ravenously on the prowl for fresh flesh.

In the early twentieth century, however, a swashbuckling adventurer named Vihjalmur Stefansson lived on an Eskimo-like diet for five years and solved the riddle for where the Inuit are finding vitamin C. After eating almost entirely chops, steaks, organ meat, fish, chicken, and fat for five years straight, Stefansson said, "If you have some fresh meat in your diet every day and don't overcook it, there will be enough C from that source alone to prevent scurvy."

Discover elaborated on Stefansson's, um, discovery and postulated that all one needs in terms of a daily intake of vitamin C is about 10 milligrams of the stuff. The magazine cited a nutritional study that determined that the Inuit diet supplied far more than that each day. A 3.55-ounce chunk of raw caribou liver, for example, contains about 24 milligrams of vitamin C, while seal brain has nearly 15 milligrams. Uncooked muktuk has 36 milligrams. Take that, scurvy!

Rather than the threat of scurvy, though, a much more immediate concern for zombies would appear to be similar to issues raised by critics of the Atkins diet. The quickest, most efficient way to create energy to fuel bodies occurs when carbohydrates are converted into glucose. When there are no carbs around, the clever little combustion engines that are our bodies turn to fat for burning and, in a pinch, can break down protein. This is at the heart, so to speak, of Atkins's approach. As *Discover* pointed out, this is why many Arctic scholars refer to the Inuit diet as the original Atkins: lots and lots of fat and protein with very little or no carbs.

Atkins detractors say that eating so much protein puts too much stress on our kidneys. And some studies have indicated that the increased levels of protein bring on an increase in the production of organic compounds called ketones, which can lead to chronic decreased kidney function. Other health risks associated with high-fat and high-cholesterol diets include ulcers, diarrhea, and, worst of all, heart problems.

So why have some researchers found that the cardiac-related deaths for the Inuit are about half of what they are for average Americans when roughly 50 percent of Inuit caloric intake comes by way of fat? At the risk of sounding like a three a.m. infomercial, it seems to be because of that old nutritional battle—good fats versus bad fats. And good fats rule! Because the Inuit eat only wild animals, they consume far less saturated fat than the highly processed junk most Americans are used to tossing down our gullets. And remember all those chilly, slithery, cold-water creatures our Arctic cousins enjoy? They are packed full of omega-3 fatty acids. Fatty acids may sound like something really bad for you, but in fact they are quite helpful in preventing heart disease and in strengthening the vascular system. Whale blubber, it turns out, is about 70 percent monounsaturated (or good) fat and 30 percent omega-3 fatty acids. It's the perfect dish!

The only problem for zombies is that they generally aren't eating whale blubber—they are eating us. So it could be argued that their health risks are considerably higher than the average Inuit's. Because if we are what we eat—and recent obesity trends, sadly, do seem to be bearing out that this is the case—and zombies are eating us, they are consuming way too much saturated fat, bad cholesterol, and, maybe worst of all, trans fats.

Trans fats, that stuff french fries are often cooked in, are created out of poor, innocent vegetable oils. The oils are transformed when an extra hydrogen atom is dolloped onto their molecular structure and they become evil, heart-crushing fats that happen to taste very delicious. Eric Dewailly, a professor of preventive medicine at Quebec's Laval University, was unequivocal in denouncing trans fats, in a conversation with *Discover*. "These man-made fats are dangerous," Dewailly said. "[They are] even worse than saturated fats." The magazine then went on to explain exactly why that is. "They not only lower high-density lipoprotein cholesterol (HDL, the 'good' cholesterol) but they also raise low-density lipoprotein cholesterol (LDL, the 'bad' cholesterol) and triglycerides"—the fat chemical. "In the process, trans fats set the stage for heart attacks because they lead to the increase of fatty buildup in artery walls."

And there you have it—too much bad fat is bad for you. Hear that, zombies? So if you are worried about a zombie's nutritional intake, there is an easy way to fix the problem: force yourself to have a healthier diet.

Of course, all of this could be for naught. At least one highly regarded zombie expert is quite certain that zombies are physiologically incapable of normal nutritional activity and are unaffected by the risks I've outlined here. "Recent evidence has once and for all discounted the theory that human flesh is the fuel for the undead," Max Brooks wrote in his groundbreaking 2003 book, *The Zombie Survival Guide: Complete Protection from the Living Dead*. Brooks goes on to say that "a zombie's digestive tract is completely dormant. The complex system that processes food, extracts nutrition, and excretes waste does not factor into a zombie's physiology. Autopsies conducted on neutralized undead have shown that their 'food' lies in its original, undigested state at all sections of the tract."

This suggests that zombies are not like Inuit at all. And if there were any lingering doubts, Brooks smashes them as if they were cantaloupe-soft skulls: "This partially chewed, rotting matter will continue to accumulate, as the zombie devours more victims, until it is forced through the anus, or literally bursts through the stomach or intestinal lining. While this more dramatic example of non-digestion is rare, hundreds of eyewitness reports have confirmed undead to have distended bellies." And here's where you might want to put down that drumstick: "One captured and dissected specimen was found to contain 211 pounds of flesh within its system! Even rarer accounts have confirmed that zombies continue to feed long after their digestive tracts have exploded from within."

Not much to add to that, really, is there? Maybe just this: Bon appétit!

Sautéed Brain with Savoy Cabbage, Finger Pieces, and a Rich Blood Sauce

■ ■ ■

Unadorned, unseasoned brain tartare ripped fresh from a human skull might be delicious, but even the best meals grow tiresome if you eat them day after undead day. In the spirit of spicing up your culinary capabilities—and maybe even your zombific life—here is a modified version of a famous brain recipe. You don't have to be a brainiac to prepare this dish, but knowing your colander from your coriander certainly helps.

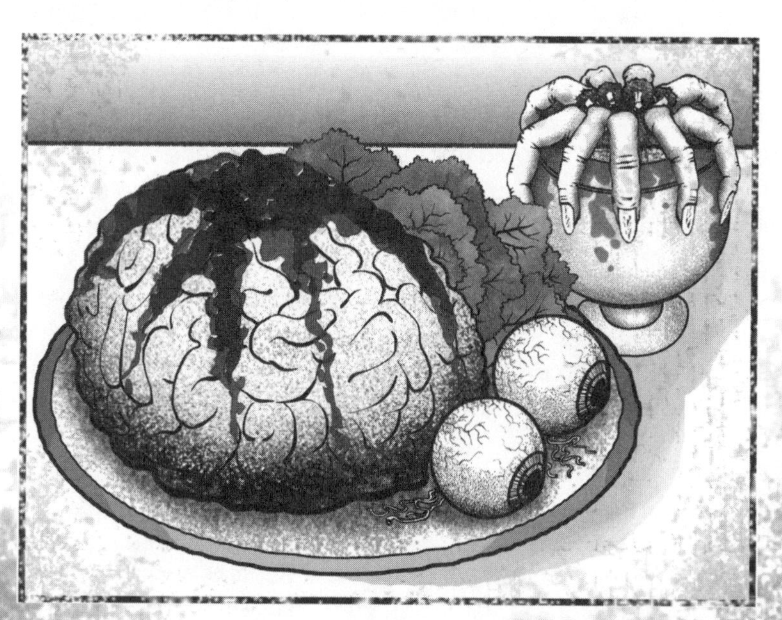

Ingredients

2 shallots, finely chopped
1 tablespoon melted human belly lard
½ cups any root vegetable, coarsely chopped
2 cups stock, human or barn animal
1 human brain (two separate lobes), cleaned*
sea salt
2–4 tablespoons dried skin crumbs
flour or ground-up bone for dredging
¼ cup melted human lard and a little extra, as needed
¾ cup finger pieces (bone and nail removed) chopped into small pieces
1 cup julienned savoy cabbage
4 ounces of the most viscous human blood
fresh ground pepper to taste
parsley for garnish

Sauté the shallots in 1 tablespoon of the belly lard until slightly mushy. Add the root vegetables and sauté for 3 to 5 minutes.

In a large pot, bring the stock to a boil. Add the sautéed vegetables to the stock. Add the brain, making sure it's completely covered in liquid (more stock may be necessary). Boil the brain for 4 minutes and remove. Allow the brain to cool.

Reduce the heat under the stock and vegetables to a simmer.

When cool enough to handle, pat the brain dry, preferably with a clean towel. Season with the salt and the dried skin crumbs and then dredge in the flour.

---

* For the most succulent dish, exercise care in preparing the brain. It's best to soak the brain in heavily salted water for at least 1½ hours—up to 3 hours if you have the patience—before you begin the dish. After a good soak, peel off the outer membrane. Feel free to pop that membrane right in your mouth, or save it for a snack if you wish.

In a heavy pan, heat up the lard and then pan sear the brain until crispy on the outside (but not too long, so that it stays tender in the middle). Remove from the pan and sauté the finger pieces with the savoy cabbage and blood.

Serve in large bowls, with the brain atop a ladleful of the finger/blood mixture. Sprinkle with salt and pepper to taste and garnish with the parsley. Serves two humans—or one really hungry zombie. Enjoy!

EARTH WORMS ARE EASY

■ ■ ■

What's really left of a dead body after a few weeks in the ground? A lot or a little?

■ ■ ■ ■ ■

IF there is one thing everybody from Winston Churchill to Groucho Marx to the Incredible Hulk to the Olsen twins can agree on, it is this: you have to have a "look." Now, in a case such as that of the Olsens, sometimes the look might not be loved by all, but it's great for branding and general recognition. Just try to think of the word *vampire* without having images of black capes, sharp fangs, widow's peaks, and Stephenie Meyer dancing in your twisted mind. The same, of course, goes for zombies. From the very first time they popped up in pop culture the undead have cultivated a look: the crumbling flesh; the moaning mouth; the awkward gait; the hollow, unblinking eyes; the toothy smirk; and the soiled trousers. You've likely seen such a look if you've ever done your grocery shopping past eleven p.m. on a Saturday night.

So how exactly did zombies get their look? Well, in a strict sense they got it by dying, being buried, and then rising from the dead. More accurately, zombies are able to look like zombies because of decomposition and the physiological breakdown of human flesh. In this chapter, let's explore what starts to happen once the heartbeat stops and how exactly decomposition helps zombies get ready for their gruesome close-ups.

When looking at the science of decomposition, you will see that some of the ways zombies are portrayed in pop culture align rather well with how things play out in real life (by which I mean death). This chapter will also go a long way toward making the existence of zombies as we know them a tad suspect. But never say never. Let's begin with the aspects of death and decomposition that support what we know about the undead, although even there we will encounter complications.

Most people agree that when your heart stops beating, you're dead. That is not always the case, of course, and folks have been known to be medically and technically dead before returning to life. In general, though, when the old ticker goes, we quickly follow. Incredibly, the process of the body's decomposing begins about four minutes later. That's less than half the time it takes to boil an egg! Or run a mile! This process has a totally cool name that is probably also the name of a German death-metal band: *autolysis*, or self-digestion.

As autolysis commences, your cells have no access to oxygen, and your bloodstream encounters more carbon dioxide and less acidity as the cells become toxic. Enzymes attack the cells, which then explode in a shower of fluids containing many of your body's remaining nutrients. This is not good. The organs containing higher numbers of enzymes or water, such as the liver, are destroyed faster. That's one reason that the liver, as well as other organs, are removed during

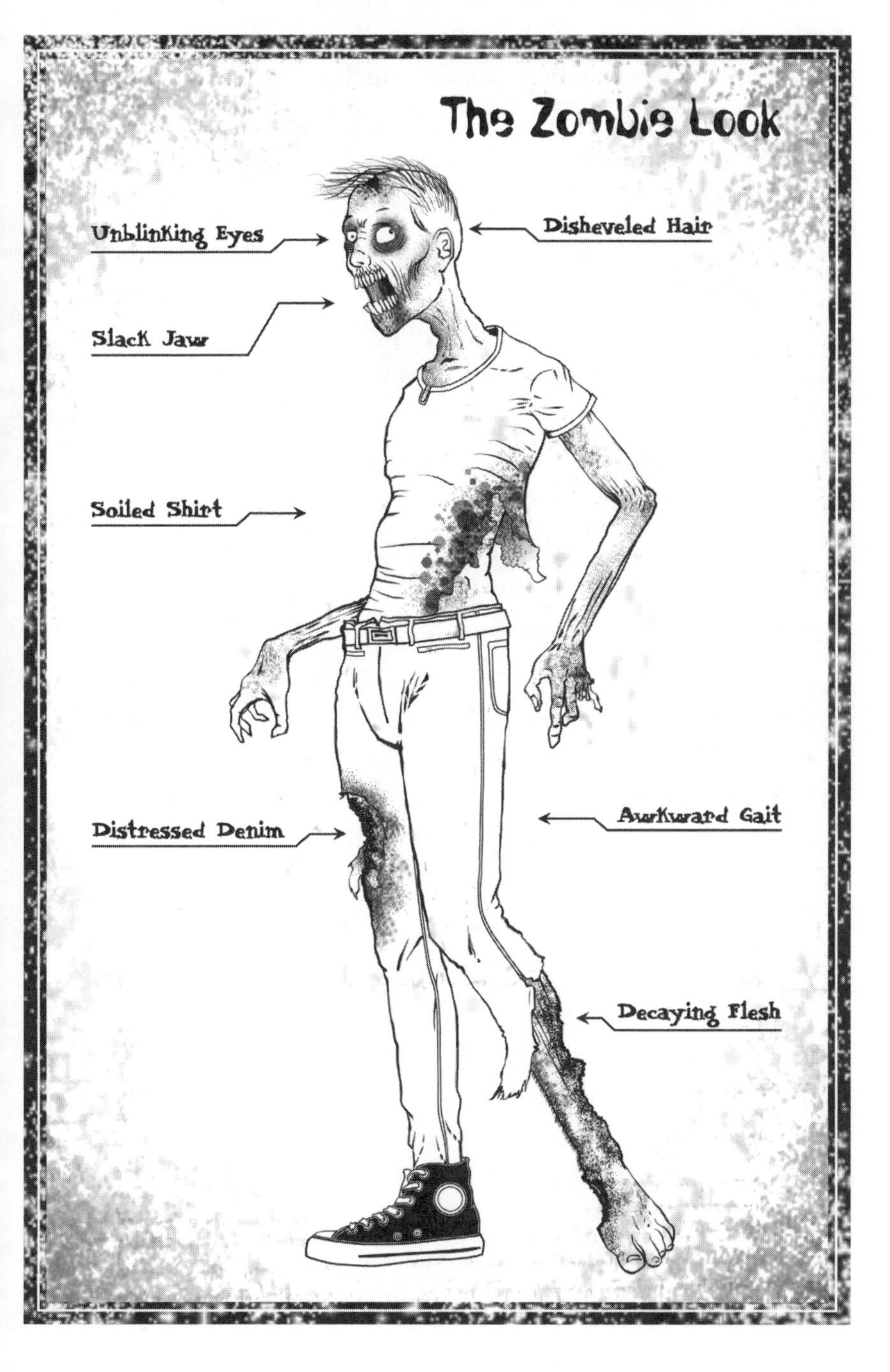
The Zombie Look
Unblinking Eyes
Disheveled Hair
Slack Jaw
Soiled Shirt
Distressed Denim
Awkward Gait
Decaying Flesh

mummification—thus preserving the body that much longer—as we will see soon enough.

While all this horrific business is taking place, your body's blood has given up in the face of gravity and raced to the regions closest to the ground. That is why you will often see a corpse's back and hindquarters—assuming the person has died while lying down—turn redder than the rest of its body, as the blood has moved away and taken its pink-tinged capillaries with it.

Amazingly, while this violence is taking part beneath the flesh, there are not many outward signs of decay until a few days have passed. Curiously, even rigor mortis—that preferred symptom of all stiff-legged, lurching zombies—takes a few hours to occur and does so only when muscle proteins coagulate. In fact, when death first strikes, a phenomenon called primary flaccidity kicks in, with the muscles relaxing entirely. This fact casts a shadow of doubt on all those zombies who, once bitten, begin immediately to wobble about as though their joints are soldered together.

Not to worry, though, because just ahead is the really excellent stuff: putrefaction. This is what Dr. Arpad A. Vass, adjunct associate professor in forensic anthropology at the University of Tennessee (yes, the same place that keeps a "body farm" in order to study decomposition), once described as "the destruction of the soft tissues of the body by the action of micro-organisms (bacteria, fungi, and protozoa) and results in the catabolism of tissue into gasses, liquids, and simple molecules."

That is—maggots! And mites! And beetles and scuttle flies and coffin flies. How many maggots? It's hard to say for sure, but in one experiment a 156-gram piece of meat was left out for twenty-four hours, and when the meat was inspected there were 48,562 maggots feasting on it. What does that mean to you and me? Maggots can devour 60 percent of a cadaver in about a week. They are extremely efficient little eaters; having posterior spiracles sure doesn't hurt. As the name suggests, where they breathe is located on their butts, so they don't need to take a feedbag break because of something as pesky as breathing.

But let's get back to that putrefaction. Intestinal bacteria really excel here. Corpses produce tons of this stuff, and it mucks up all of the tissues and blood vessels and bloats the body and turns it funny colors like green and purple and black. Incidentally, two of the main chemicals created at this stage of death are the aptly named putrescine and cadaverine.

It is here that I would like to turn over the description of what a dead body looks like at this point to Dr. Kenneth V. Iserson, author of what many people consider to be the decomposition bible, *Death to Dust*. In that impressive tome, Dr. Iserson writes:

> By seven days after death, most of the body is discolored and giant blood-tinged putrid blisters begin to appear. The skin loosens and any pressure causes the top layer to come off in large sheets (skin slip). As the internal organs and fatty tissues decay, they produce large quantities of foul-smelling gas. By the second week after death, the abdomen, scrotum, breasts, and tongue swell; the eyes bulge out. A bloody fluid seeps out of the mouth and nose. After three to four

> weeks, the hair, nails, and teeth loosen and the grossly swollen internal organs begin to rupture and eventually liquefy.

Many variables affect the rate and intensity of decomposition, such as temperature and the amount of moisture in the air, but scientists who have studied the art of the afterlife have a pretty good idea what comes next. Dr. Vass, in fact, has such a good idea that he even devised a formula to express the process: $y = 1{,}285/x$. In this equation, y is the time in days it takes for a body to become skeletal after death, and x is the average centigrade temperature over that time. So with hotter temperatures, where bacteria and other decomposing agents thrive, you have a shorter period of decay.

Vass is quick to point out, however, that just because someone is a skeleton does not mean that they are done decomposing (as anyone who has ever seen *Evil Dead II* can tell you). The good doctor notes that:

> The skeleton also has a decomposition rate that is based on the loss of organic (collagen) and inorganic components. Some of the inorganic components we use to determine the length of time since death include calcium, potassium, and magnesium. As with soft tissue, these leach out of bone at a rate determined primarily by temperature and exposure to moisture.

According to Vass's research, within a year after death, bones often have algae or moss growing on them. Several years later, a tree or shrub's roots might find purchase in the cavity of a bone. Let's pause for a minute to consider what all of this means for our good friends the zombies. There's a lot to digest, but one thing you almost never see is a zombie with a tree growing out of its head, so those fearsome flesh-chewers haunting the multiplex, we can deduce, have been in the ground for under a decade. But if they expired more than a year before crawling from their graves—as seems likely with the toxic-rain-brought-on-by-mysterious-gases-produced zombies in *The Return of the Living Dead*, to take but one example—then wouldn't they appear more skeletal than they do? The short answer is yes. The long answer, as you probably have already guessed, is longer.

In *Zombie CSU*, writer Jonathan Maberry looks at the Romero model for zombified behavior and concludes that "[zombies] are the recent dead brought back to some semblance of life, but they are definitely corpses." "Recent dead" would explain why they don't look like skeletons with azaleas blooming out of their bloomers. But if, as Maberry suggests, Romero's zombies—and subsequently many other people's zombies, too—are newly departed, then we are still going to bump up against some scientific inconsistencies. Rigor mortis typically doesn't set in until about three hours after death, brought on when the lack of oxygen interrupts the production of adenosine triphosphate, which confuses the easily confused calcium ions, and they end up bonding with the wrong proteins like cheap skanks. (Adenosine triphosphate is a cellular enzyme that uses oxygen to direct calcium where it needs to go, such as to one's bones. No more oxygen means no more adenosine triphosphate, and that means the calcium ions are adrift and no longer able to strengthen bones.)

This lasts only about a day and a half, when the enzymes begin chewing apart the muscle tissue and rigor mortis dissipates.

Assuming that the Romero-ian zombies' stiff-jointed shuffle is a byproduct of rigor mortis—or whatever passes for rigor mortis among the undead—we are talking about a fairly finite window for reanimation. The dead, it would seem, are rejoining the living sometime between three and thirty-six hours after expiring. It certainly makes sense that it wouldn't be much longer than that. Once we begin to approach a week past the great expiration notice, we enter the physiological territory that, as Dr. Iserson notes, involves "large sheets" of skin slipping off the corpse at the slightest touch, like snow off a slanted rooftop.

I can't think of a single instance when zombies were quite so delicate that they were liable to lose a limb if they shook hands with their prey a little too roughly. Sure, there are scenes where it *appeared* that was possible. A few that leap to mind involve our three heroes from the original *Dawn of the Dead* darting through a mall packed with undead shoppers. The ease with which Stephen, Peter, and Roger knocked the villains to the linoleum floor was comical. Those zombies were, for the most part, incredibly weak, tumbling as a result of not much more than a spirited shove. And yet, at no time did their ghoulish flesh simply slough off, as would be the case with a typical week-old corpse. Given all of that, it seems logical to assume that the zombies of *DOTD*, as well as the zombies from many other pop culture productions, have been dead for somewhere between two days (when severe rigor mortis would have worn off) and a week (when their skin would start shedding in a strong breeze).

Another possibility is articulated in *Zombie CSU*. Maberry writes, "Are zombies truly dead, partly alive, or something else? Personally I'm leaning toward 'something else,' some state between clinical death and actual life." Some state, indeed. Sounds like Maberry is onto something, but what? What is this mysterious "something else"? Ah, but that is material for another book, I'm afraid—the metaphysics of zombies. For now, let's stick with the task at hand.

So far we've only discussed what happens to an average person's unadulterated body after death and what that means for the zombies we know and love. It is, however, incredibly rare for a body to decompose without being treated in some way. Generally in American society, this means a body is embalmed to help preserve it as long as possible—or until the maggots get to it, in any case. The practice of embalming is practically as old as death itself. Dr. Iserson notes how early solutions used for this purpose included "essential oils, aloe, salt, myrrh, precious spices, honey, plaster, wax, and sugar." (Also the recipe, oddly, for a surprisingly tasty sponge cake!) Arguably the most famous embalmers of all time were ancient Egyptians, who began the practice way back in 4400 B.C. and were so good at it that the Smithsonian keeps an exhibit of mummification practices on heavy rotation.

Dr. Vass notes in his paper for the University of Tennessee a couple of examples of unintentional embalming, which if taken up by zombies would likely change their iconic look considerably. In one of the cases, a woman's body was discovered by workers who were clearing brush near a road. The woman was fully clothed and her corpse had undergone zero decomposition. There was no evident gnawing done by insects.

At first the inspectors assigned to the case reflected on these facts and determined that she was the victim of homicide and that it had taken place not long ago at all. Upon further examination, however, they realized that the woman died about four months before she was discovered. Apparently, her killer had covered her with insecticide to cover the stench associated with a decomposing body. This worked as a giant liquid shield against any mites or beetles or flies or, yes, maggots that might have a hankering for her flesh. Even weirder—an autopsy showed comparatively minimal internal decomposition, as well. Investigators deduced that the insecticide had penetrated the woman's lungs and other internal organs.

As for the second example of unintentional embalming, Dr. Vass writes about the case many years ago when grave robbers discovered the body of a former Civil War colonel. According to Vass's paper, the corpse was still in remarkably good shape and even still in possession of flesh on its bones. How in the world was *that* possible? The colonel was of course a former VIP, and back in the day such folks were treated well even in death—the coffin containing the high-ranking chap was made entirely out of lead. The coffin, Vass writes, "had sterilized the body by poisoning the microflora and decomposition had not progressed past initial autolysis."

If zombies ever got hip to this lead coffin thing, one can only imagine how they'd lobby to bring them back. Both the well-made casket and the bug spray would significantly improve the way zombies look and would also allow them to linger underground a bit longer before popping out. They'd certainly be less anxious about getting down to business before that whole day-seven-skin-flaking-off condition kicked in.

Were the undead truly interested in preserving their good looks—and teeth—they would take a page out of the Egyptian playbook. Nobody spent hard-earned cash in insane quantities on corpses like the Egyptians did. And thanks to contemporary advancements in computer imaging (MRIs, CT scans), radiology, and forensic science, we now have a pretty good idea how they did it. Earlier we talked about how organs with high levels of enzymes or a lot of water were more susceptible to deterioration. This was understood by Egyptians, too, who found a way around this hurdle by cleverly yanking those parts out. A 1994 article from the *Journal of the Royal Society of Medicine* describes the practice this way: "A four-sided plate covered the abdominal left flank incision where the liver, stomach, lungs, and intestines had been removed (the heart was usually left in place during mummification). The viscera were subsequently wrapped and replaced in the thorax, which was typical of the embalming procedure of that period."

In that same article, the description of mummification and its result on the corpse could stand in for a physical description of a very attractive zombie: "The soft tissue was more radio-opaque than normal tissue. This may be explained through the use of Natron (a natural balm) being used as a dehydration agent and also the application of various resins to the skin during the mummification process. The ears appeared to be well preserved although the nose appeared displaced."

The aspect of mummifying that I believe zombies would appreciate most has to do with chewing. I'm talking about those long, white, essential tools for any undead guy or gal who hopes to satisfy their flesh jones. During a series of tests in 2005 at Massachusetts General Hospital, researchers observed a mummy to have remarkably powerful teeth still in place. The old fella's three left molars were there—as were incisors, canines, and bilateral premolars. Some deterioration was detected in the mummy's left central incisor, but that was likely due to dental disease or postmortem trauma. The message to zombies could not be clearer: brush twice a day, floss regularly, and get mummified, and you, too, could have far healthier teeth to help you tirelessly and effectively devour brains.

In the end, mummification turns out to be too good to be true for zombies. What I failed to add to the list of removed organs just now was one that is not only crucial but, thanks again to recent technological developments, is widely believed to have been extracted during mummification. That's right—the brain! Sorry zombies, maybe the Egyptians weren't so cool after all. Seems the brain is all too likely to swell and liquefy after death and is far too attractive to destruction-minded enzymes. For the Egyptians, this meant it had to go. The only question for them was: How? Warning: if you are a zombie or just a lover of brains, you may want to skip the following paragraph.

The brain was removed through the nasal cavity, of course! A hooked metal rod was shoved through to smash up the dead brain into tiny pieces that could then be pulled out bit by bit. This is problematic for zombies on a fundamental level. If all dead people had their brains yanked out by way of their schnozzes, it sure would make it tougher to subsequently rise from the grave and function in any way that approaches what zombies are used to. As we've demonstrated earlier in the book, not only does it appear that zombies have working brains, but they have fairly highly developed brains capable of intense and nuanced emotions such as anger, yearning, rage—and anger. (Okay, maybe they aren't that nuanced, but there can be no arguing that at least the zombie brain does exist.) Or, as Gertrude Stein might have put it, when it comes to the cerebral cortex of the undead, there is certainly some there there. How else to explain zombies' ability to walk, identify victims, and decide not to clean up after themselves?

As it turns out, maybe mummification is not the answer for preserving and maintaining the zombie lifestyle past the cruelly quick one week that Dr. Iserson gives them before their faces, arms, legs, and other parts essentially melt away as if they were nothing but *Raiders of the Lost Ark* extras. Perhaps they'd be better off just shellacking themselves in bug spray and curling up inside a solid lead coffin.

But the truth is that one of the wonderful things about zombies is that they don't seem preoccupied with the gloriously superficial stuff that takes up so much consideration for the less enlightened masses we call the living. Zombies don't really care about how their hair looks (matted), what kind of clothes they are wearing (shredded), or what sort of car they are driving (they're not). They only really care about eating human flesh. And besides, just as with Winston Churchill, Groucho Marx, the Incredible Hulk, and the Olsen twins, zombies have something better than flawless skin—they've got a look.

And there is one other thing that zombies don't have to worry about that we do: what to do with the dead bodies.

It's morbid and complicated stuff, but it clearly must be done. To borrow yet again from Dr. Iserson (and this time he, in turn, was borrowing from a contemporary sociologist), the fact that the dead must be dealt with and dealt with swiftly is as implacably true as death. Quoting the scientist in *Death to Dust*, Iserson writes, "The physicality of a human corpse is undeniable. It is a carcass, with a predisposition to decay, to become noisome, obnoxious to the senses, and harrowing to the emotions. Disposal of such perishable remains is imperative."

When put like that it almost sounds like a description of zombies. Noisome, obnoxious, and harrowing to the emotions—yes, yes, and yes! Not that there's anything wrong with that.

FOUR

SEX AND THE SINGLE ZOMBIE

■ ■ ■

The undead are far from the only organisms that reproduce asexually, but are they the scariest?

■ ■ ■ ■ ■

THE title of this chapter, as you may already know, borrows from the 1962 book *Sex and the Single Girl*, written by former *Cosmpolitan* magazine editor Helen Gurley Brown. That timely effort observed a liberated, independent-minded class of women flooding the workforce, primed for the inchoate sexual revolution. What follows in these pages is a chronicle of an equally fascinating phenomenon—only instead of bed-hopping Betties on the make, we will turn our collective sex-obsessed eye toward zombies and their decidedly less lusty reproductive habits.

Just like tiny bacteria and other single-celled organisms and plants, the undead reproduce asexually. While one could argue quite well that what ghouls are up to is far closer to the spreading of a virus (as we'll explore in chapter 6) than to sex, it is nonetheless true that they do reproduce but without the fertilization of an egg. In other words, they reproduce asexually. The most common form of zombified procreating comes when zombies take a nibble out of a heretofore perfectly unzombified organism (read: you) and in so doing pass on the monstrous virus or somehow impart the scary DNA to their victims. Of course, sometimes it just takes a stinky, mysterious toxic leak (see *Return of the Living Dead*), an unseen, curious satellite disaster (see *Night of the Living Dead*), or a possessed hand (see *Idle Hands*—no really, see it!).

In this chapter, we'll examine which plant or single-celled organism zombies most resemble in their asexuality. We will try to determine what, if any, benefits such creatures enjoy as a result of this decidedly unsexy reconstituting. And, conversely, we'll examine what zombies, bacteria, fungi, and, especially, the freshwater snails found in New Zealand lakes and known to friends as *Potamopyrgus antipodarum* are missing out on by not getting down. Because, as many of those snails know better than any of us, good things often come in twos.

Let's begin with some good news for asexuals. As this type of reproduction requires only one (female) organism, it is generally considered a faster and more efficient—if not nearly as exciting—way to multiply. Of course, this does not really do much for zombies who actually rely on a second organism—us!—to help them procreate. The process of zombifying, however, almost always takes less than nine months, so at least part of what is true for other asexual beings is also true for the undead: reproduction is faster.

Plunging right into the different forms of asexual reproduction, let's see if we can find one that seems especially zombielike. For a process that is supposedly a quick way to increase numbers, let me just say that asexual reproduction

is pretty damn complex. All of that DNA mutating and cell forming is a bit tricky, as it turns out.

One fairly simple and prominent form of not getting it on is vegetative reproduction, which is a common way for plants to reproduce. Vegetative reproduction is a type of asexual reproduction that takes place free of germ cells—there is no egg, no sperm, no fertilization per se at work. In this instance, part of the plant breaks or shoots off from the original organism and a new individual arrives with DNA identical to that of the mother plant. This process can involve the production of a fresh plant growing out of an existing organism's rhizomes, or roots. Some flowers and fruits, such as the strawberry, reproduce this way. Very few multicellular beings are able to engage in vegetative reproduction, but one that comes close to being able to do so is the starfish. When the arm—or leg—of that little devil snaps off, it is able to regrow the remaining parts of itself with DNA that mirrors the DNA of the original starfish. This would be as if a zombie had its arm lopped off and then that arm somehow grew the rest of its body up around it. Anyone who has seen *Re-Animator* is familiar with zombie body parts twitching with life after they've been severed; what is far less familiar is the idea that the part could then become a whole.

A close—but not that close—cousin to vegetative reproduction is budding, wherein cells simply split apart and form new individuals. Budding is more common among single-celled organisms but can also take place among multicellular creatures. These creatures include the microscopic freshwater-swimming hydra and certain types of jellyfish. Amazingly, the hydra reproduces in a way that resembles that of certain plants, with the new organism growing out of a tiny pocket in the body cavity of the hydra. And much of what the momma hydra eats goes toward the new, budding offspring. Soon enough, extremely small tentacles begin to bloom on the bud. Eventually the bulging addition to the hydra's body breaks free of the mothership to roam the world (or at least the pail of water it's splashing about in) on its own.

Binary fission might sound like something a villain in a James Bond movie develops to blow up the world, but it is really just a process of asexual reproduction favored by protists, bacteria, and other simpletons. You may recall observing binary fission in action when you gazed down the business end of a microscope during eighth-grade biology class. The bacteria formed a second identical DNA strand, which elongated to the point that the new strand pulled away until one became two. (Celine Dion, you can have that metaphor for free. You're welcome.)

By now you are probably wondering about the poor algae. Well, wonder no more: algae of course have their own favored form of asexuality, and they call it spore formation. One way of thinking about this is that it involves mitosis and mitospores, but I prefer to think about it as basically another fancy way to say that cells are split, new organisms are formed with DNA that matches the mother algae, and absolutely zero sex has taken place. It is a rather unglamorous way to reproduce, which is why fungi are fans of it and why it shares at least some qualities with zombie behavior.

The last form of asexual conjugating we will run through for now is fragmentation. As you might guess, this is what it is called when a new organism grows out of a fragment of a still existing individual. Fungi and plants like to do this. So do some very small single-celled worms.

It has been believed by biologists for some time that there are great and significant benefits to reproducing sexually as opposed to asexually. Less understood is exactly how those benefits manifest themselves, though genetic diversity sure is a big plus. At first blush, in fact, it can seem as if there are quite meaningful advantages to asexual procreating. Speed and efficiency, which were mentioned earlier, should not be minimized. Let's say there are 100 tiny organisms in one lake, and 50 of them are male and 50 female. And right next to them in another lake there are another 100 of the same organisms, only the breakdown there is 1 male and 99 females. If the females in both lakes are asexual reproducers, then it is clear that the one with 99 females will populate the lake at a far faster rate than the one with an even split—the rate, in fact, would be nearly double.

So why don't asexuals—or zombies—eventually take over the world? That is the sort of question Maurine Neiman, assistant professor of biology in the University of Iowa College of Liberal Arts and Sciences, sought to explain when she traveled not long ago to study the freshwater snail species of *Potamopyrgus antipodarum*, or *P. antipodarum*, in the lakes of New Zealand.

What Neiman found there was downright spooky in its similarities to zombie culture. One of her school advisers had previously studied the snail populations of New Zealand, so Neiman was aware of a tiny parasitic worm sharing the waters with the snails. In a paper published in 2009 in the journal *Molecular Biology and Evolution*, Neiman referred to these worms as the "sterilizing trematode parasite, *Microphallus sp.*" For our purposes we will just call them parasites. And what these parasites were able to do was both fascinating and terrifying. First they would feast on our friends the snails, but not just take a tiny bite or two. No, the parasites would swarm the snails by the thousands, completely occupying the cells, effectively castrating the snails and killing them. But the parasites didn't stop there. The killer worms really had their tiny, beady eyes on a bigger fish—or in this case a duck that also inhabits the New Zealand lakes. The ducks, you see, were fond of snacking on snails, too. The parasites, clever buggers that they are, figured this out and also figured out that the best way to get to the ducks was by moving into the snails. "It turns out the parasites are keeping the asexuals at bay by killing them off," Neiman told me. "Without them, the asexuals would take over." To sum up: the asexuals, the snails, are killed off when the duck eats the snail, a development that the parasite has orchestrated.

An additional point worth noting is that parasites are drawn more often to organisms they can easily recognize. Since asexuals—as we saw with the plants—pass their entire genomic sequence on to the new organisms they create, the parasites have grown more familiar with the asexuals than the sexuals, who create brand new, original DNA with every offspring. "So if you are rare or a novelty, that gives you an advantage," Neiman said.

There is another advantage to reproducing sexually, and this one is Neiman's major breakthrough to date. The New Zealand snails are among the small batch of creatures that engage in both sexual and asexual behavior. This allowed

Neiman to monitor and track both forms of reproduction in the very same species in the very same environment. Historically, biologists have observed that often if there is an asexual and a sexual version of the same species, the asexual individual has lived in a harsher environment (which, in turn, would make it more difficult to find a partner anyway), and this has led some researchers to chalk that up as a reason asexuals have not come to dominate the globe.

Having both types of reproducers in one place enabled Neiman to hone in on another reason sexual organisms have an advantage over asexuals. What Neiman observed is something she calls mutation accumulation, a crucial point in understanding the genomic superiority of sexuals over asexuals. "Mutation is a random process," Neiman wrote to me in an e-mail. "And when random things happen to proteins that are functioning well, they will usually make the protein function less well, or not function at all."

Neiman compares this to trying to fix a television by hitting it with a hammer: "Good mutations that happen to co-occur in a genome with bad mutations are unlikely to spread when the genome is transmitted asexually, since the good mutations are permanently linked to bad mutations. This is akin to 'throwing the baby out with the bathwater.'"

Examples of harmful mutations, Neiman explained, "are the single mutations that cause Tay-Sachs, color blindness, Huntington's, dwarfism, and certain breast cancers, or mutations associated with traits with a multi-genic basis like myopia, diabetes, and Alzheimer's."

Because sexual creatures produce offspring with unique DNA—and also reduce the chances of passing along harmful mutations by having two distinct sets of genes play a part in reproduction—they are not at risk to accumulate mutations to a fatal degree. Or, as Neiman put it to me, "this means that in sexuals, mutations can be selected independently of their genetic background—something that isn't possible in asexuals."

In summing up her work with the *P. antipodarum* freshwater snails of New Zealand, Neiman told a science Web site, "This is the first study to compare mutation accumulation in a species where sexual individuals and asexual individuals regularly coexist, and thus provides the most direct evidence to date that sex helps to counter the accumulation of harmful mutations." Huzzah! Eat it, asexuals!

Of course it could be argued that since zombies do not pass on an identical genomic sequence when they "reproduce," they are not in danger in terms of mutation accumulation in the same way that the snails are. That could be argued, but who is to say with absolute confidence that this is not exactly what is going on when a dark-eyed ghoul sinks his venomous incisors into your bicep? Perhaps at that moment he is transmitting his DNA code—harmful mutations and all—directly to you. Has that ever been disproved? For now, let's not worry our pretty, sexually reproducing little heads over these mere technicalities. Instead, let's look at some other ways in which sex is not just fun—but can save your life.

As a *Forbes* magazine story noted not long ago, "Having regular and enthusiastic sex . . . confers a host of measurable physiological advantages, be you male or female." It also noted a definitive study performed by the *British Medical Journal* in the mid-1990s claiming that "men [with] the highest frequency of orgasm enjoyed a death rate half that of the laggards." Of course, zombies are not so much concerned with death rates—as they are already dead—but were they to engage in the physical act of love, there are, as *Forbes* observed, other considerable health benefits. Let's run through them.

- There is a large postcoital increase in the production of the hormone prolactin. This hormone prompts the creation of new neurons in the part of the brain that enhances one's sense of smell. That will make the stopping and smelling of roses in life that much easier.
- You will have a stronger heart. How much stronger? If you are a man and you have sex three times a week or more (I know, I know, but stay with me), your chances of suffering a heart attack or a stroke are cut in half.
- More lovemaking equals more testosterone, which equals more robust muscles and bones. Another equation to keep in mind: one super boot-bumping session can burn some 200 calories, or about 6 ounces of brain for fellow zombies.
- Ladies, this one is for you: A 2002 study of nearly three hundred women found that women who were getting regular action were less likely to suffer from depression. Semen contains a hormone called prostaglandin, which can affect female hormones after lodging itself in the genital tract. Get laid and get happy!
- Oxytocin should not be confused with Oxycontin. The latter is the pain medicine to which Rush Limbaugh once admitted he was addicted. The former is a hormone that spikes upward by five times its normal amount just

before orgasm. On its own it is not so impressive, but when loads of oxytocin is released, it triggers endorphins to shoot up, and those babies can be a great pain reliever. Zombies are often in pain, what with the groaning and all the shovels to the head, and this could be a way for them to ameliorate future discomfort.

- A study done by a Pennsylvania university found that folks who have sex a couple of times a week produce 30 percent more immunoglobulin A, an antibody that can help ward off colds and flus by strengthening one's immune system. This actually could be really helpful for zombies who are, traditionally, rather susceptible to contracting viruses.
- Something else zombies would probably want to improve, for purely utilitarian reasons, is the strength of their teeth. And great news: semen possesses zinc and calcium, both of which can prevent teeth from rotting. Of course, this point is not about sexual intercourse but instead about the benefits of oral sex. That counts as sex, too, you know!
- Last, but certainly not least, frequent ejaculations, some urologists argue, can decrease the risk of contracting prostate cancer. This argument seems a tad sketchy, but the gist of it goes like this: In order to manufacture semen, the prostate and other parts of the body extract citric acid, potassium, and zinc from the bloodstream. At the same time, if there are any carcinogens—the nasty stuff that can cause cancer—in the bloodstream, those get sucked into the process, too, and then in turn are shot out of the body. *Forbes* noted that the

> reduction in cancer risk associated with this physiological production could be offset by maintaining multiple sex partners, who, it's been estimated, can increase a fella's chances of encountering the Big C by 40 percent.

There you have it. So why don't more—or really, any—zombies seek solace in the warm embrace of a love partner? Well, actually they do. At least they do in the uniquely twisted zom-com directed by Peter Jackson—that lovable, lunatic New Zealander best known as the man who made *The Lord of the Rings* movies—*Dead Alive*. Lovemaking, zombie and otherwise, figures most prominently in the narrative.

The story opens with Kiwi adventurers transporting an exotic new breed of animal back from 1950s Sumatra to New Zealand in order to research the creature. We learn right away that the animal is the progeny of rat and monkey mating on Skull Island. To be precise, the rats have "raped" the monkeys and born out of this unholy coupling is something known as—the rat-monkey! (This is the first instance in the movie when sexual reproduction is not really so cool.) During the course of relocating the beast to New Zealand, the vermin finds its way to the lovely Wellington Zoo, where it sinks its viral fangs into our protagonist Lionel's overbearing mother's arm. A handful of manic scenes later, mom is in full-on zombie mode and hungry for whatever human flesh she can get her gummy mouth on: her nurse, the local priest, and even a pack of random cemetery hooligans with questionable pompadours.

Lionel, trying desperately to hold onto the house after his mom's perceived passing, secures the new zombies in the basement while trying to figure out how to proceed. Not easy moving on in life when you have a cellar full of the undead. Things get even dicier for our unlikely hero when his loathsome uncle Les comes sniffing around, looking for a cut of the will. It is at that most inopportune time that a quite rare act of zombie sex occurs, mostly off screen, but rather audible nonetheless. While Lionel deals with Les, the priest and the nurse go at it behind the dining room door. After Les leaves, Lionel walks in on them mid-gruesome thrust and pulls them apart, but not before the nurse is able to gnaw off half of the good father's face.

The gestation period for zombies seems to be considerably shorter than it is for us humans, because before long Lionel discovers the zombie baby, wailing and screeching inside an old-fashioned radio. Raising the brat is beyond a nightmare, as the little nipper is as hungry as a human baby—only hungry for flesh! Lionel puts the baby in a brand-new pram and takes him for a stroll in the park, but he has to bolt from the park in a hurry after Junior escapes from the carriage and attempts to eat a little girl playing peacefully on the grass. "Hyperactive," Lionel explains to the other (extremely nervous) mothers. And how!

The zombie baby is an example of how not all sexual reproduction is superior to and healthier than asexual recombinations. That's something zombies might want to consider if, as Maurine Neiman and others have documented in considerable research, sexual reproduction has immediate and ongoing advantages over asexual reproduction. If, however, the only offspring that zombies can create are anything like the baby in *Dead Alive*, then Neiman's theory gets all shot to hell. And then bitten by a rabid infant. And kicked in the crotch by the same beastly babe utilizing the leg of some unfortunate soul as a kind of human cudgel.

The lesson we should all learn is this: sexual reproduction, while an effective and at times zesty means of copulating, is not always best for every species. It seems to be the more robust, ultimately more effective way for the snail species *Potamopyrgus antipodarum* found in freshwater lakes of New Zealand to reproduce. But, interestingly, it is not so hot for a couple of other species found in New Zealand, namely the rabid, eye-bulging, vein-chomping offspring of a viral rat and monkey, and the face-crushing, nut-walloping zombie baby of a ghoulish priest and a nurse. Good to know.

FIVE

UNSAFE AT ANY SPEED

■ ■ ■

Why zombies walk (and stalk) before they run.

■ ■ ■ ■ ■

LET'S say you are spending a long weekend in a remote farmhouse, far from crowded city streets, other people, organized law enforcement, or anyone else who might conceivably be able to help on the off chance that something terrible happens. You are there with a handful of friends, and you are under attack by zombies. They are marching, en masse, out of the woods to gnaw on your limbs, to eat your brain, to zombify you. There are, what, 50 or 75 or 250? It's hard to count just how many, but they are coming for you, and you are terrified.

Quick: What's the first thing you notice about them? Probably their hygiene, which is terrible, of course. Ravaged and pocked and chunky, like a sweaty teenage boy on a pizza bender. But, hey, it's not easy being undead.

And the second thing you notice? Their gait. It's. So. Amazingly. Slow. Like, superslow. It's more of a rhythmic shuffle than an actual walk. How does any zombie in its right mind think it's going to catch a victim at that pace? "Let's go, fellas, pick it up a little!" you want to yell at them, but then you remember that if they were faster they would have devoured you by now. Eventually they *will* eat your brains. It might take them a while, but they'll get around to it, and when they do, when they are poised to chomp, just above the ear, into the side of your head with those oddly sharp incisors, the last thought you will have is this: why do zombies walk so slowly, anyway?

That is a good and, as it turns out, complicated question. Because while historically zombies have moved at a rather meandering rate—and they certainly did in early films such as *White Zombie*, *I Walked with a Zombie*, and George Romero's classic oeuvre—lately they've begun to hustle for their supper. A gathering of zombies no longer resembles Boca Raton during the early bird special. They can really book. In more recent movies, like Danny Boyle's terrifying dystopian vision, *28 Days Later*, and, curiously, a remake of Romero's *Dawn of the Dead*, the undead have been unleashed. Zombie purists will say these newfangled freaks are all wrong. They will decry this quickening as unnecessary meddling. Some hardcore zombie-ists will even say science is on their side. That's because in the early 1980s, a Harvard enthobotanist named Wade Davis traveled to Haiti to study voodoo culture and religion. What he learned—and later published in a bestselling book and eventual movie called *The Serpent and the Rainbow*—suggests that zombies should definitely walk slowly—if they're able to walk at all.

Davis returned from his first trip to Haiti carrying several samples of something he called zombie powder. The main

0
28
223
8

ingredient in the mixture was an incredibly toxic substance found in a handful of underwater animals—most notably puffer fish, those deceptively cute, prickly little swimmers with physiques like Dizzy Gillespie's cheeks. The poisonous substance found in puffer fish livers and reproductive organs is tetrodotoxin. As an April 1988 issue of *Science* magazine noted, Davis said the stuff is so lethal (100 times more poisonous than cyanide) that 3.5 grams of it could render a 160-pound person "comatose." That's exactly why the puffer fish poison was so attractive to Haitian criminals looking to make a fast buck by selling off zombie slaves. Davis said he saw these crooks in action—feeding the poison to victims until their metabolism and heart rate slowed to the point that they appeared dead. The poor souls were then buried alive, dug up soon after by the would-be zombifiers and treated again with hallucinogens and more tetrodotoxin. And voilá, zombie slaves were born! Or, rather, reborn! Undead!

Some existing research makes Davis's claims sound downright plausible. In 1982, a Columbia Presbyterian Hospital pathologist tested a sample of Davis's zombie powder on rats and a rhesus monkey. The pathologist later wrote to Davis that the rats "appeared comatose and showed no response at all to external stimuli. The electroencephalograph continued to monitor central nervous system activity, and the hearts were not affected. Certain rats [and the monkey] remained immobilized for 24 hours and then recovered with no apparent sign of injury." Is it any wonder, then, that the etymology of the word *zombie* can be traced to the Haitian Creole term for someone who has shuffled off his mortal coil, only to shuffle it back on again—*zonbi*?

When you ingest tetrodotoxin (not recommended), it sticks itself to your sodium ions and doesn't let the ions travel along nerve cell membranes to deliver vital electrical messages from your brain to your muscle fibers. This means that the signals your brain is trying to send to your muscles to move this way or that—to, say, walk—never reach their destination. Symptoms include a deteriorating ability to use one's limbs properly, muscle contraction, an extreme drop in blood pressure, and, often, death. In any event, you soon become immobilized. That makes ambling tough; running would be out of the question.

Zombie purists, arguing the case for unhurried horror, might point to Davis's findings and say that of course zombies walk slowly. Hell, they're lucky to be alive! But not so fast: much of Davis's science was later questioned by several experts on toxins, including C. Y. Kao, working out of the State University of New York system, who was a "well-known authority on tetrodotoxin." How much of an authority? He once made his own tetrodotoxin analyzing kit! That might sound strange until you remember the Japanese delicacy, fugu fish, which is a type of puffer fish and also contains tetrodotoxin. Kao didn't do it just for zombie kicks; he had gourmands in mind, as well. One day Kao ran his own tests on Davis's zombie powder by separating its chemical components and concluded that there was such a small amount of tetrodotoxin found that its role as the "causal agent in the initial zombification process is without factual foundation." Harsh!

Following Kao's findings, there were many scientific volleys between Davis and his detractors. Much of the debate centered on how much tetrodotoxin is the right amount if the goal is to make someone close to dead but not actually dead. It was agreed upon that roughly 70 grams of the supertoxic solution might do the trick. Davis seized on this as proof that what he had described was possible. He admitted that to zombify in the way he had seen was laborious, delicate work, and it wasn't for everyone. "I've never maintained there is some kind of assembly line producing zombies in Haiti," Davis told *Science*.

Indeed, even Davis himself was initially skeptical about the existence of real zombies (silly man). He first heard about the idea one rainy evening on a visit to the Manhattan home of Dr. Nathan S. Kline, whom Davis describes in *The Serpent and the Rainbow* as a "psychiatrist and pioneer in the field of psychopharmacology." Dr. Kline, it seems, had made many trips to Haiti in an effort to secure a sample of zombie powder but had had no luck. Enter Davis, whose reputation as an adventuring scientist preceded him.

That evening at Dr. Kline's apartment, however, Davis remained unconvinced that zombies were real, and he was resistant to doing Kline's bidding. The story of one unfortunate soul changed all that. This was the tale of a forty-year-old Haitian peasant named Clairvius Narcisse.

On April 30, 1962, a man by that name was admitted to the Albert Schweitzer Hospital in the Artibonite Valley. He had a high fever, was extremely achy, and possessed "general malaise." He also happened to be spitting blood, which is never a good sign. On May 2 at 1:15 p.m. Narcisse was pronounced dead. The following day he was buried near his home village of L'Estere. His remains were believed to be under the large concrete memorial his family had placed upon his grave site.

Yet eighteen years after his apparent death, a fellow claiming to be Clairvius Narcisse introduced himself to his sister, Angelina Narcisse, at an outdoor market in L'Estere. Angelina had no reason to doubt that the person before her insisting he was her brother was anyone other than Clairvius Narcisse—he had used his childhood nickname known only to a handful of family members. Clairvius told his sister that he had refused to sell off his portion of inherited land to his brother and so his brother did what any vengeful sibling might do and hired a *boko*, or sorcerer, to zombify Clairvius. His grave had been robbed and Clairvius was beaten so badly that he'd become completely disoriented. For two years he was enslaved with other zombies in northern Haiti. Eventually, the zombie sorcerer was killed, and Clairvius spent the ensuing sixteen years fretting and wandering, afraid to go anywhere near his brother. When he finally heard that his brother had died, he felt it was safe enough to reenter L'Estere.

It was the case of Clairvius Narcisse—and the attendant media attention—that convinced Wade Davis to travel to Haiti in search of zombie powder. Lovers of zombie lore are lucky that he did.

But back to the matter at hand: the rumored assembly line producing zombies in Haiti. Should those hollow-eyed victims survive their abduction and tetrodotoxin injection, the models produced would most certainly be of the slow-moving, Romero-ian variety. And it was because of Romero's films that for many years we couldn't even comprehend an alternative. As the Web magazine Slate noted a few years back, there is a scene in Romero's zombie masterpiece *Night of the Living Dead* that says plenty about the speed of flesh eaters. In a televised news update on the menace, we see a reporter interviewing the police chief. The intrepid muckraker zeroes in on the most important piece of information, asking, "Are they slow-moving, chief?" The cop seems incredulous, replying simply, "Yeah, they're dead."

Ten years later, for Romero's wicked follow-up to *NOTLD*, *Dawn of the Dead*, the movie's tagline read, "When there's no more room in hell, the dead will walk the Earth." Walk the Earth. Not sprint, jog, or even saunter the Earth. Walk. So George is pretty sure that his brain suckers are moving at the right speed.

Then why in recent years have zombies sped up? One interesting theory that's been floated by certain zombie-o-philes—okay, me—is that in this multitasking, techy, globally connected, hurrying world we now live in, even zombies have to figure out how to keep up. It's zombievolution—just as the human attention span has shrunk, the better to follow the endless crawl of info at the foot of so many screens, so have zombies been required to step a little more lively. The world is faster now—and so are the undead.

Ultimately, it may be next to impossible to say with absolute conviction that, scientifically, zombies are meant to move at a leisurely pace. In fact, science comes down on the side of those favoring faster-footed undead in at least one regard. How's that? Well, assuming that when a zombie attacks us, a certain number of us fight back, it would seem reasonable to assume that fear is thick in the air at that moment. There is even a condition for what I am getting at: fight or flight.

At the moment when a conscious being assesses a threat and decides which path to choose, an involved series of physiological responses kicks in. Even for a zombie who has little doubt about fighting, rather than fleeing, there is a very good chance that in the heat of the moment its body undergoes the same changes you or I do when it's time to throw down. Namely, hormones such as adrenaline and cortisol are released into the bloodstream. This has several effects: the heart rate increases, the digestive tract slows, and blood flow ramps up in the primary muscle groups. All of which has an immediate impact on your strength and energy levels—they spike upward.

What does this mean for a zombie grappling for brains? It means the dang thing is pumping adrenaline like nobody's business, and its muscles are straining like Jack LaLanne on a push-up spree. The zombie in question is going to be able to haul just a little faster. Or, depending on how afraid or threatened the brute is feeling, maybe a lot.

History—and serious zombie purists—would suggest that this is an anomalous situation, that the undead plod rather than plow, but history has been wrong before. Happily, we can hang on to one crucial fact that no matter where you stand on the question of speed you cannot deny. Zombies are real. Why else, as *The Encyclopedia of Death and Dying* tells us, is zombification listed as a crime in the Haitian Penal Code? Check article 246 if you don't believe it. As the encyclopedia's authors note:

> Local beliefs about body, mind and spirit recognize a separation of the *corps cadavre* (physical body) with

> its *gwo-bon anj* (animating principle) from the *ti-bon anj* (agency, awareness and memory). In zombification, the latter is retained by the sorcerer (*boko*), usually in a fastened bottle or earthenware jar where it is known as the zombie astral. The *boko* either extracts it through sorcery, which leaves the victim apparently dead, or else captures it after natural death before it has gone too far from the body. The animated body remains without will or agency as the *zombi cadavre*, which becomes the slave of the *boko* and works secretly on his land or is sold to another *boko* for the same purpose.

So there. But then what happens when all of these animated bodies start turning on their masters and begin chomping on their arms to infect them? Is a zombie apocalypse even really possible? Let's find out.

The Seven-Day Zombie Workout

Ever wonder how the undead maintain their endurance and manage to stay so lean and strong? With a superstrict workout regime with absolutely zero days of rest, of course! Follow this fitness schedule and in no time you'll be ripped enough to hold down an adult-sized human struggling for his life long enough to get a good angle on the juiciest artery. If you feel your commitment to a healthy, ghoulish lifestyle wavering, just remember: no pain, no brain!

Day One

- Triceps: Dig out of grave by reaching hands above head, grabbing a fistful of dirt, and pushing upward while splaying fingers as if reaching for the sinister moon (or like a *Thriller* video backup dancer).
- Lats: After slurping out the innards of an innocent passerby, lift the drained-of-organs carcass above your head and toss to the side in a ridiculously menacing fashion. Repeat.
- Chest: When captured and placed in a holding cell, press both hands against the wall and push with all your might. You can't possibly escape that way, but your pecs will soon be bulging inside your tattered T-shirt.

Day Two

- 27 reps of smashing arms through faux-concrete walls (each arm).
- 412 laps—walk up the stairs in a mall; take the escalator down. Avoid fountains and the food court.
- 16 reps of getting knocked to the ground by a shotgun blast to the shoulder. Struggle to your feet. Feel the burn in the lower abs.
- Cool down with a human heart-and-blood orange smoothie—high in protein and fiber.

Day Three

- Cardio day: Walk endlessly in a lurching manner through a shadowy field, preferably in the English countryside or Connecticut. A zombie of average weight, 175 lbs, walking at 3 mph for 4 hours burns 1,386 calories! You'll be skinny-jeans ready before you can say: unnnhhhhh!

Day Four

- Attack a dorm full of unsuspecting college coeds. Mangle victims thoroughly for maximum increase of muscle density.

- While lumbering along with your fellow zombie hordes, bump into as many of them as possible—great upper leg exercise.
- Crash through ceiling while pursuing prey. This helps shed calories from huge dorm-based meal. Looks cool.

Day Five

- Focus on abs: Hang upside down from a streetlamp on a deserted boulevard near the governor's mansion. When you suddenly hear the roaring blades of a SWAT Team helicopter approach, pull yourself up to confront the invasion, while raging and waving your fists to indicate your displeasure. When SWAT Team fires automatic weaponry at you, dodge this way and that to avoid taking a bullet in the head. For faster development of abdominal muscles, rage harder.

Day Six

- While chasing a banker, fleeing and shrieking like a little schoolgirl, try to increase your speed to catch him. Zombie purists will decry the fact that you are moving faster than the undead did in *Night of the Living Dead*, but that won't matter to you when the pounds are just melting off.
- When the same banker attempts to elude you by leaping into a lake, follow and try to swim. You likely don't know how to swim, which does wonders for the cardio benefits of the workout.
- Cool down at bottom of lake.

Day Seven

- While gripping your victim from behind in a terrifying bear hug, thrust your hands deep into his chest cavity, grip rib cage with each hand, and tear apart by extending your elbows out to your sides as if you are about to break into the Funky Chicken dance.

- Grab your victim's vital organs and squeeze —liver, spleen, etc.—15 times. Start now and Popeye forearms will be yours by summer, just in time for paparazzi shots at premiere of blockbuster zombie flick.
- Insert thumbs in dinner's eye sockets. Pull face apart to create space for slurping out brain.
- Enter standard pushup position with face over victim's brains. Lower your body slowly toward the floor. Slurp some brain. Slowly push your body off the floor. Repeat 25 times or until all brain is consumed.

SIX

WHAT THEY DON'T TEACH YOU IN HEALTH CLASS

...

How fast and how far will a zombie infection spread? Is a zombie apocalypse truly possible? The science of the infected.

■ ■ ■ ■ ■

THE nastiness of some viruses and illnesses can be limited by getting proper rest, drinking plenty of fluids, covering your mouth when you cough, and keeping clean. Back during the great H1N1 (or swine flu, if you prefer) scare of 2009, no lesser of a human being than our great and mighty president gave a weary nation the same counsel my late grandfather gave me every time I entered his immaculate home: go wash your hands! And it actually can help a little.

Slopping soap on one's hands, however, doesn't do much to ward off the infection spread by zombies. Drinking juice doesn't do a hell of a lot, either. There seems, in fact, to be very little in terms of preventive medicine that we humans can take to stay healthy once the zombie virus starts a-spreadin' (Will Smith's excellent and exhaustive search for a vaccine in *I Am Legend* notwithstanding).

Unlike with mumps, measles, chicken pox, and the clap, among other historically virulent illnesses, we've yet to develop a vaccine that might be taken to prevent the telltale symptoms of zombiedom from inhabiting our bodies: the restless, stiff-legged wandering; the eyes ringed with dark circles; the skin-dripping visage; the withered ability to drop bons mots at cocktail parties.

What all of this suggests, as you may have guessed by now, is this: we're totally screwed. Once the undead virus is unleashed, it would seem there is really nothing that can be done to leash it back up. At that point the question is no longer will we survive, but rather, how long can we survive? How long before the infection contaminates each and every one of us, hopping ponds, rivers, and oceans with a lusty desire to do its sickening business?

That's where math comes in. Unlike psychics, math is really good at predicting the future—or at least making some sense of the future. Perhaps if we know exactly how long we have, if we know what the score is, then maybe, just maybe, that knowledge will help motivate us to find a cure. Okay, probably not, but when the zombies come, math might be the best hope we have. So let's have a look at the numbers.

Amazingly, there are actually numbers to look at. That's because in 2009, a small group of highly motivated Canadian mathematicians and researchers took up the humanitarian cause of modeling an infectious zombie outbreak. The results, naturally, were not pretty. But they sure were definitive!

Philip Munz, Ioan Hudea, Joe Imad, and Robert J. Smith? (yes, the question mark is part of his name, and, yes, he's entitled to it because he helped model a zombie outbreak) worked out of the School of Mathematics and Statistics at Carleton University. They published their results in a paper

S=SUSCEPTIBLE
Z=ZOMBIE
R=REMOVED

that was included in the book *Infectious Disease Modelling Research Progress.*

Early in the paper, the authors note the inherent obstacles facing their task. For starters, the origins of a zombie virus can be difficult to nail down. They are often mysterious and uncertain and are not triggered by the same genetic contortions as, say, the influenza pandemic of 1918 (more on that later). A zombie outbreak can, as the Canadians note, derive from the vaguely described radiation of *Night of the Living Dead*, or from an airborne virus as in *Resident Evil*, or a sinister government serum (*28 Days Later*, *I Am Legend*), or mad cow disease (*Zombieland*). All of this mystery makes modeling a little tricky, even for math.

What Mr. Smith? and his team are sure of is that "when a susceptible individual is bitten by a zombie, it leaves an open wound. The wound created by the zombie has zombie saliva in and around it. This bodily fluid mixes with the blood, thus infecting the (previously susceptible) individual." Sure, okay.

Our academics have their cause of illness. Next they needed to agree on what type of zombie they were dealing with—specifically, fast or slow? You purists out there will be glad to hear that they went with the Romero-ian model, the lumbering variety of brain taster.

The basic equation for determining the rate of a zombie outbreak contained three variables: Susceptibles, Zombies, and Removed. Those first two categories should be clear enough, and Removed simply refers to humans who had died by way of a zombie attack or otherwise.

Here's where things get complicated. As we all know, just because one is dead does not mean one cannot later become undead. The paper's authors dealt with this phenomenon by allotting a special parameter to those who have been made undead. An additional parameter was assigned to Susceptibles who had been transformed into Zombies.

In an effort to minimize variables, our fearless mathletes assume that the birth rate will remain constant and thus will not influence the findings. They note that Zombies can also be made Removed by—all together now—"removing the head or destroying the brain." Well—yeah.

After addressing the question of the various parameters, the authors write, "This model is slightly more complicated than the basic SIR models that usually characterise infectious diseases, because this model has two mass-action transmissions, which leads to having more than one nonlinear term in the model."

That's putting it nicely. The truth is that the Carleton University modeling is so complicated that it cannot be rendered by the keyboard of a non-math professor level of human. Trust me, though; this kind of works out in our favor. I can skip a whole bunch of boring old chicken scratches and get right to the gory, good stuff. Yes, the findings!

In order to get there, I should first let you know that there were a couple of other factors taken into account by the researchers. For one thing, citing the great work of Mr. Max Brooks, the authors state that roughly twenty-four hours elapses between the time a Susceptible is first nibbled and its transformation into a new Zombie. This alters the modeling yet more as "Susceptibles first move to an infected class once infected and remain there for some period of time," and also, "infected individuals can still die a 'natural' death before becoming a zombie, otherwise they become a zombie."

Finally, the Canadians ran essentially three variations on their findings: one wherein a somewhat effective system of quarantining has been established; another wherein a modestly effective treatment has been developed—but while the cure can reverse the effects of zombification, it does not guarantee that the cured will stay cured; and a model accounting for the fact that we will fight back and kill our share of the undead during a series of calculated attacks.

The results are chilling. If no action is taken, the zombie outbreak will overtake a city of five hundred thousand people in roughly four days. For those keeping score at home, that's only about as long as the director's cut of the movie *Gandhi*. Speaking of the peaceful leader, when Dr. Smith? and his gang attempted a system of quarantining the undead, they determined that quarantining would only slow down the eventual zombification of everyone, and only by an additional four days.

When a treatment or cure for the virus was included in the modeling, the results indicated that, in the words of the report, "humans are not eradicated, but only exist in low numbers." Um—yay?

Finally, we come to the model incorporating a series of counterstrikes on the zombies by humans. That is, a model based on more or less every movie ever made about these particular ghouls. Unlike every movie ever made, however, the Carleton paper suggests that humans can actually win and reduce the number of zombies—and thus the spread of infection—at a rate that's downright zombielike. According to their figures, it would take only two and a half days for us to eliminate a quarter of the zombie population. And that rate of destruction would play out over time, so that after five days half the zombies would be toast and after ten days—*ten short days*—we would be living in a paradise that is 100 percent zombie free.

You don't believe it? Talk to the math, man! Because as the Canadian researchers write in their paper,

> Only sufficiently frequent attacks, with increasing force, will result in the eradication, assuming the available resources can be mustered in time.
>
> Furthermore, these results assumed that the timescale of the outbreak was short, so that the natural birth and death rates could be ignored. If the timescale of the outbreak increases, then the result is the doomsday scenario: an outbreak of zombies will result in the collapse of civilisation, with every human infected, or dead.

Before you become too dismayed, though, let's consider a few things. While the research done at that fancy Ottawa academy is certainly impressive, it is far from the final word on the speed and destructive force of a zombie viral outbreak.

As the good folks at the online Zombie Research Society pointed out in a 2009 bulletin, the Centers for Disease Control has recently beefed up its ability to respond to and deal with widespread health scares and diseases. In the interest of public safety, the ZRS checked in with a CDC official for some highly acronymic tough talk about the likelihood of a massive zombie virus epidemic or even pandemic. Similar to what the Carleton professors did, the CDC looked at four areas when it assessed the scenarios: susceptibility, exposure, infection, and recovery. The CDC determined that such a disease would come up against what experts like to call a little situation having to do with exposure. Or, as the CDC representative put it to the ZRS: "Even assuming the entire world is susceptible, if zombiism is only spread through a bite or some other close bodily contact, then you have an obvious exposure problem. And despite what movies like *28 Days Later* suggest, the faster the infection spreads the less likely it is to impact a large population."

Interesting. Additionally, the folks at the ZRS reminded me that Jonathan Maberry, in his book *Zombie CSU*, determined that almost all representations of zombies have overstated how fast and how furiously the disease could spread. In speaking with a former hospital administrator, Maberry pressed the issue—could a zombie virus move through the human population at a rate comparable to the lightning-quick infections seen in a film like *28 Days Later*, or even the twenty-four-hour pace of the classic zombie infection found in earlier masterworks such as *Night of the Living Dead*? This is the response from Maberry's expert: "Not going to happen. I could buy a reduced metabolic rate and some organ shutdown, which means I could almost buy the *Night of the Living Dead* zombie with some medical exceptions. At a stretch I could make a case for it; but the other plague doesn't follow pathogenic spread patterns."

Comforted? Not so fast. Maberry notes that if the spread of infection is slower than it generally is portrayed in pop culture, this could actually work against us humans. "If the plague does not spread as quickly as it does in more recent films," he writes, "then there is actually greater chance of it spreading farther before it's detected."

Maberry's experts agree. In discussing incubation periods, neurologist Peter Lukacs says, "Even the common cold takes 2–3 days before making someone sick. For every patient admitted, there would be several infected patients still at large. And in this modern day, people can travel to just about any part of the globe within 24 hours."

Damn you, modern day! It gets worse: Maberry's former hospital worker admits that given a longer incubation time on the disease, "we could see a frightening pattern emerging. Not an aggressive attack like in *28 Days Later*, but a more quiet and insidious attack, like we saw with the spread of HIV."

Are you comforted now? Okay, let's hang on a sec before we all go screaming into the streets because the end is so obviously nigh. No matter what we determine about the probability of a zombie plague infecting the entire world's population, it is also worth keeping in mind that there are ways other than quarantining, curing, or killing zombies to disrupt the seemingly inevitable march of zombification. I like to call these ways the *Day of the Dead* techniques, for it is in that organ-slurping installment of Mr. George Romero's unrivaled series that we are introduced to two curious ways of dealing with the unique disease.

The first method involves amputation. Toward the end of the film we see the female protagonist's (tough hottie Sarah's) boyfriend (insane military dude) receive an unfortunate chomp to the left forearm. (You may recall that an ersatz military operation has been set up in a remote bunker very much like the one that served as Dick Cheney's mailing address for eight-plus years, and scientific tests are being conducted in an effort to find some way of surviving what seems like imminent doom.) Acting incredibly swiftly and smartly, Sarah grabs her fella's machete and hacks off his arm at the bicep, above the point of infection. She then immediately cauterizes the stumpy wound with a torch that is conveniently nearby.

From there things turn a tad murky. It is not clear if the amputation has stopped the disease from spreading through the remainder of the body. The ambiguity arises because although the insane dude does not become fully zombified, we do see him shivering under a thin blanket in a manner that suggests he has either been infected and is slowly turning or someone has been screwing with the bunker's thermostat. Later, he slips out of his bed and takes the giant service elevator up and out of the bunker, in the process giving himself up as a snack to the waiting legion of zombies and also letting them ride down into the safe basement where the rest of the humans reside.

So either this guy was crazy like a fox and knew exactly what he was up to—three of the four heroes of the film make it out alive, while all of the bad guys get chomped—or he was gradually succumbing to the infection and was doing his fellow zombies a solid. In either scenario, though, what is fairly certain is this: amputating an appendage can reduce the rate of infection within a given population, a parameter not factored into the great modeling work done by our friends to the north.

Another, more nuanced, approach to dealing with the plague was also introduced in this same film. That has to do with the socializing efforts undertaken by a character facetiously named Dr. Frankenstein by all the hotheads in the bunker. He may have been a little screwy, but by about halfway into the flick he sure seems to be making progress in his efforts to reteach zombies how to comport themselves in a civilized society (the irony being that many of the humans here act like monsters). "We're lost unless we can make them behave," the doctor tells his dozen or so colleagues in *Day of the Dead*, pleading for more time to work with his prize pupil, Bub, a rather tall zombie.

Dr. Frankenstein makes a good case for this approach. Before getting iced by his own boss, the doctor helps Bub unlock an array of foggy, pre-zombie memories. At different points we see Bub recall how to use a telephone; how to put on headphones, the better to enjoy Beethoven; and how—poignantly—to fire a gun. He is beginning to communicate just as all hell, and then Bub, breaks loose. His appetite for human flesh persists, but the film indicates that with enough retraining Bub may even lose the taste. This incremental socializing, it follows, is potentially another technique for not only slowing, but perhaps eventually nullifying the spread of a zombie virus.

But if we are going to find theoretical models in one film, then we must honor others. Just because it is horrifying to imagine does not mean we can turn our backs on scenarios such as the ones presented in *I Am Legend* or *28 Days* (or even *Weeks*) *Later*. Given the severity and destructive capacity of the diseases depicted in those films, it might be helpful, in thinking about how society could contend with that level of heartbreak, to look to the past for an approximate comparable example of a virulent worldwide pandemic.

There is one virus that can rival zombie fever when it comes to mysteriousness, persistence, virulence, and sheer unadulterated evil. I'm talking, of course, about the confusingly named Spanish influenza of 1918. It's a confusing name since it is unlikely that the flu originated in Spain, and it actually lasted until 1920. What is known is that it spread across the globe at a rate and force never before or since seen in a virus—in just over a year's time—and in three rioting waves of illness, the flu killed some 50 million people and infected about 500 million more, or roughly one-third of the world's population in 1918. And—zombielike—the sickness curiously struck young people most harshly; about 50 percent of the inflicted were between the ages of twenty and forty.

It wasn't until just recently—2006 to be exact—that scientists were able to make a good guess about the genetic makeup of the 1918 flu strain. Using the frozen tissues of a victim, researchers at the Centers for Disease Control, working with a team from the Armed Forces Institute of Pathology, re-created the entire genomic sequence of this flu. In a CDC press release about the findings, it is noted that "the pandemic's most striking feature was its unusually high death rate among otherwise healthy people aged 15–34." Now that sounds pretty familiar, oui?

Given the size and destructive nature of the pandemic, it is not at all an overstatement to call the illness the mother of all pandemics. In fact, that's exactly what two scientists from the Armed Forces Institute of Pathology called it in a

2006 paper for the CDC. Jeffery K. Taubenberger and David M. Morens call it that for a good reason. It seems that every flu-based pandemic since the one from 1918 has been related to the Spanish influenza in its genetic coding. Or, using the word of the authors, all other flu pandemics are "descendants" of the 1918 scourge.

The scientists observe that many of the nastier proteins embedded in the 1918 virus developed out of the bird flu just before the pandemic, and many subsequent flus are "composed of key genes from the 1918 virus." Oddly, the H5N1 bird flu that rocked the aughts is not one of them. "On the contrary," Taubenberger and Morens write, "the 1918 virus appears to be an avian-like influenza virus derived in toto from an unknown source, as its 8 genome segments are substantially different from contemporary avian influenza genes."

An unknown source? Why don't you just call it what it is, guys—zombies! And it gets weirder still.

> For example, the 1918 nucleoprotein (NP) gene sequence is similar to that of viruses found in wild birds at the amino acid level but very divergent at the nucleotide level, which suggests considerable evolutionary distance between sources of the 1918 NP and of currently sequenced NP genes in wild bird strains.

After sifting through the data, the authors arrive at a decidedly murky place: "All of these findings beg the question: where did the 1918 virus come from?"

They first point to solid evidence that the flu may have come from the same place as later pandemics, such as one in

1957 and another in 1968—namely, originating in Eurasian bird flus—and then decide that to pinpoint exactly where the 1918 virus began would require human samples from before the pandemic struck and after. And those, rather conveniently, do not appear to be available.

In the end, the Spanish influenza and a zombie plague have much in common. Maybe the thing that binds them most closely is the uncertainty that surrounds each outbreak.

In the event of an undead spread, most experts agree that it would be nearly impossible to locate patient zero. And with the 1918 situation, during over ninety years of research and analysis, we aren't even much closer to being able to nail down a country of origin, much less the first infected body. What we do know about each disease could pretty much fit into a single paragraph. Two, tops.

We now know the genomic sequence of the Spanish flu, though not that of a possible zombie virus. We can guess that the 1918 illness's reach was aided by the fact that soldiers returning home from World War I carried the flu with them. Could *World War Z* provide a similar boost to the zombifiying of a people? Debatable. The earlier influenza, as Taubenberger and Morens detail, flourishes in places with "lower environmental and human nasal temperatures, optimal humidity, increased crowding indoors, and imperfect ventilation due to closed windows and suboptimal airflow." While it is difficult to contend with absolute assuredness that this is also the case with an undead disease, it was definitely true in *Dawn of the Dead*, as anyone who has spent much time in a crowded shopping mall knows all too well.

In terms of how fast and how far both viruses can go if not treated, we know that the Spanish pandemic struck all over the globe over three successive waves that lasted not much more than twelve months in total. For zombies, it is a tad more difficult to calculate, but that's certainly not for lack of Canadian effort. Dr. Smith? and his gang figured that a zombie plague left untreated would completely infect a city of five hundred thousand people in about four days. Extrapolating from that—and incorporating absolutely zero new parameters—we find that it would take about 2,472 days, or 6 years, 9 months, and change, for the entire United States (population 309 million) to be done in. And it would require a whopping 54,400 days for the z-flu to strike everyone in the world. That works out to a cool 149 years and 15 days.

What does all of this back-of-the-envelope doodling mean to you and me? It means a few things:

1. As noted earlier, we are screwed.
2. We are screwed, but maybe, *just maybe*, we are not screwed for several decades, so get out there, have fun, live it up, learn a second language or how to play guitar.
3. See point number 1.

Ultimately, what all this comes down to is what researchers Niall P. A. S. Johnson and Juergen Mueller decided in a 2002 article titled "Updating the Accounts: Global Mortality of the 1918–1920 'Spanish' Influenza Pandemic." It is their work that many academics credit with establishing the most agreed upon figure for the 1918 virus's death toll: 50 million people. These two are no slouches. It is instructive and meaningful, therefore, when they write, "One of the most difficult problems for those working on past outbreaks of disease is that of data: what data there are tend to be inconsistent and of questionable validity, accuracy, and robustness."

In other words, all of the mathematical modeling in the world, although extremely enlightening and often eye opening, cannot change this one simple fact about the coming zombie plague: we are . . . in trouble.

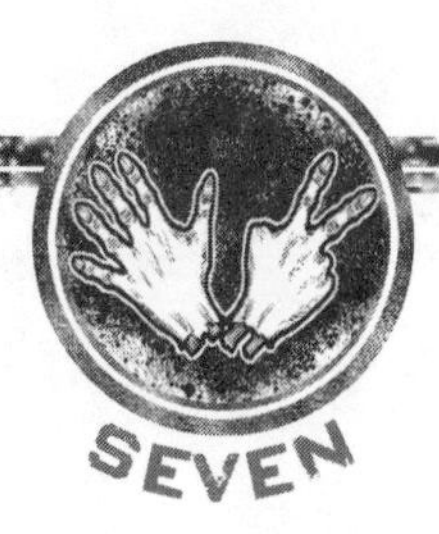

DO ZOMBIES DREAM OF UNDEAD SHEEP?

■ ■ ■

Not only are zombies undead, they are massively unslept. An exploration into the science of sleep.

■ ■ ■ ■ ■

HITTING the hay. Catching some z's. Crashing out. No matter how you refer to it, I think we can all agree that sleep is incredibly important. It's so important that if you don't get enough of it—if you are not able to rest your body and mind—you may start to act irritable, sullen, and depressed. You will begin to feel sluggish, and it will grow harder to think through even simple daily situations like what side of your bread you like buttered. If things get really bad, as we will see, you may even be driven to the point of madness. And in the absolute worst cases, people with sleep disorders have been known to rise out of bed and, while remaining completely asleep, commit cold-blooded murder. Yes, these humans kill much like a zombie kills, unemotionally and with little or no memory of the event. This is ironic considering how zombies get as much sleep as they do sex—in other words, almost none at all.

Curiously, during sleep the human brain operates similarly to a zombie brain, as we explored in chapter 1. Looked at in this way, in fact, it could be said that humans are never more zombielike than when they are asleep (except during Black Friday, of course). The interesting thing about that is what it suggests: humans become more zombified during the act of sleeping but they also begin to resemble zombies physically if they do not get enough rest. There is much to explore here!

All of this becomes even curiouser—and rather creepy—when one stops to consider just how prevalent sleep disorders are in this country. According to the Centers for Disease Control, in 2008 nearly 70 percent of Americans surveyed said they had experienced some level of insufficient sleep over the thirty days prior to when the poll was conducted. And about 28 percent had had rough nights during at least half that time, while 11 percent revealed that they felt unrested *every single day of the month.*

Putting it conservatively, more than a tenth of the population is walking around right this minute in a sleep-deprived fog resembling the shuffling state of zombies, and many more among us are not far behind.

There is interesting and often surprising research related to sleep deprivation and the brain. Throw in some zombies and, well, things get even more interesting. For example, the prefrontal cortex, which as we've seen is the area of the brain that is responsible for complex cognitive behaviors such as decision making, adept social behavior, and expression of personality, is more active in a sleep-deprived person. Conversely, the temporal lobe of the brain (the part that helps with language) becomes less active in people who are sleep deprived. In other words: zombies are not so good at small talk, but they sure know what they want and how to get it!

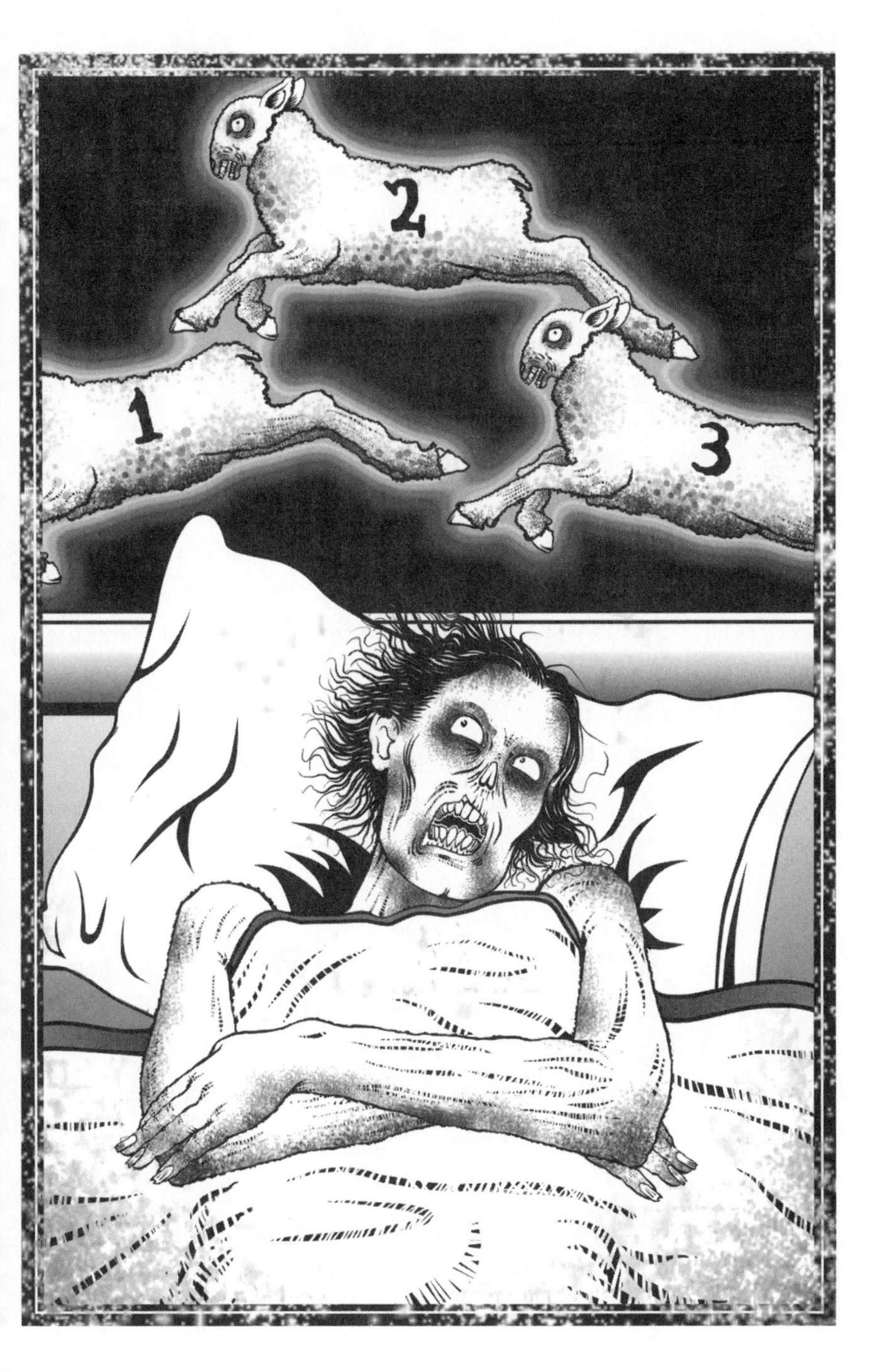
2
1
3

A 1996 study undertaken by researchers June J. Pilcher and Allen I. Huffcutt of Bradley University's department of psychology explored three main areas concerning sleep deprivation: the effects on cognitive performance, motor performance, and mood. For this chapter, we can think about these as thinking, driving, and laughing. In filing their report, Pilcher and Huffcutt studied the sleep patterns of 1,932 subjects and involved some 143 coefficients. To get a deep understanding of how those who underslept respond to their situations, Pilcher and Huffcutt put the subjects through a series of tests with strict time limits and conditions. For example, cognitive performance tasks like math problems were administered for between six and ten minutes, and motor performance tasks such as walking on a treadmill were performed for between three and eight minutes. And dare I say, the results were startling!

The authors wrote, "Partial sleep deprivation appeared to have a considerably greater overall impact on subjects than either short-term or long-term deprivation. In terms of type of measure, sleep deprivation in general appeared to have the least effect on motor tasks, a greater effect on cognitive tasks and an even greater effect on mood."

Let's break it down. While most researchers agree that it is only after about forty-five hours that sleep deprivation seriously begins to erode our motor and cognitive skills and mood, Pilcher and Huffcutt posit that partial deprivation of less than forty-five hours can have serious and immediate impacts on us in those areas. Moreover, they suggest that our moods and our abilities to perform mental tasks suffer to a greater degree during partial sleep deprivation than does our ability to act physically.

These findings just happen to align practically perfectly with the narrative gestalt of the movie *Zombieland*. For it was in Garland, Texas, you'll recall, that our nebbishy, springy-haired hero, Columbus (named for his hometown), played by the nebbishy, springy-haired actor Jesse Eisenberg, finds himself with an incredibly sleep-deprived zombie in his lap.

At least that's one reading of the film. In a flashback sequence, Columbus reminds us that two months earlier "patient zero took a bite of a contaminated burger at the Gas 'n Sip." Right around that time, Columbus was on a three-week World of Warcraft, pizza, and Code Red Mountain Dew bender, during which time he had barely seen the light of day. And for good reason—the light of day is darkening.

One night there is a frantic banging at his door. "I don't usually unlock my door for the sounds of panic, but my neighbor, 406, is insanely hot." So he invites her in, offers her a mug of Code Red Mountain Dew, and when she wearily explains that a homeless man recently tried to take a bite out of her, he overlooks this small inconvenience, as he's always dreamed of "brushing a girl's hair over her ear." Soon enough he has his chance when 406 asks, "Do you mind if I just close my eyes for a minute?"

Had Columbus read Pilcher and Huffcutt's report, he would have known the detrimental effects that partial sleep deprivation can have, especially on cognition and mood, while motor skills get off relatively easy. Then he might not have been so surprised when one short dissolve later the hottie from 406 wakes up in full mucus-spewing, twitching, stiff-jointed, face-chewed glory. Two blows to the head with a heavy ceramic toilet lid and 406 is not only decidedly less hot—she's dead. Unfortunately for Columbus, mad cow disease "became mad person, became mad zombie." There is a national epidemic brewing, and it is going to take some serious balls and a delicious, scenery-gnawing performance by Woody Harrelson to combat it.

From the *Zombieland* example we can see that there are clear and certain dangers when we are dealing with a zombie suffering from partial sleep deprivation. Columbus's neighbor conks out, rests for too short a time, and awakes in the throes of a high-grade zombie freak-out. But what if 406 had slept longer? What if she were not so much deprived of sleep as she was fully rested? That is, let's suppose zombies were capable of getting regular and normal time in the sack—say between seven and nine hours a night. What would that do to the zombie brain and what, in turn, would zombies to do us?

Based on a 2002 article by J. Allan Hobson and Edward F. Pace-Schott of the Laboratory of Neurophysiology at Harvard Medical School, the short answers respectively are: a lot and nothing good. The two professors set out to answer at least a couple of questions: Can the neurobiology of sleep help us understand the neural basis of conscious experience? And does sleep have consequences for cognitive function such as learning and memory? They sought to address these questions and more by observing neuronal activity during a series of sleep experiments and at different points in the circadian rhythm, from waking to sleep to deep rapid eye movement, or REM (no relation to Michael Stipe), sleep to non-REM, or NREM, sleep. By utilizing a piece of wicked-cool technology called a Nightcap, basically a shower cap–like hat that can monitor the brain activity of a sleeping subject, Hobson and

Pace-Schott could determine what was going on—and in what part of the brain—during both lighter NREM sleep and the dream state of REM.

The first general observation made from these experiments was that "there can no longer be any doubt that NREM and REM sleep support quantitatively different states of consciousness." That sure sounds cool, but what does it mean? With a supercomputer hooked up to a sleeper wearing a Nightcap, the researchers could practically see into the subject's brain to identify when they were having thoughts and when they were producing something closer to the hallucinations that we call dreams. Hobson and Pace-Schott continue:

> So waking suppresses hallucinosis in favour of thought, and REM sleep releases hallucinosis at the expense of thought. This contrast in mental activity corresponds to shifts in the activation pattern from waking to REM at both the molecular/cellular and brain-regional levels. We propose this correlation represents a deep causality: as the brain goes, so goes the mind.

That, my friends, is mind blowing! The deeper we go into sleep, the more our brain function is actually altered to resemble a nonthinking, hallucinating creature. A zombie deep in sleep is even more zombie than normal.

Neurologically speaking, in deep, or REM, sleep, the Hobson and Pace-Schott paper notes: "Different regions are . . . hyper-activated (the amygdala, paralimbic cortices and certain multimodal association areas) and deactivated (the dorsolateral prefrontal cortex)." As our friend Dr. Steven Schlozman has noted (and as we discussed in chapter 1)—zombies can be seen as little more than walking amygdalas, as that is the part of the brain that deals with and understands emotion and emotional responses. So, again, deep-sleeping zombies, it follows, would essentially be strengthening their zombie muscles while they doze. Hobson and Pace-Schott additionally point out that the amygdala transmits feelings of anxiety, which is why so many of us have dreams wherein we are soaring to dangerous heights or naked in front of strangers. It doesn't take too much imagination to see the undead as being extremely anxious, to the point of irritability, about where their next meal of flesh and bones is going to come from.

The Harvard fellas' paper also observes that the dorsolateral prefrontal cortex (DPC) is "deactivated" during REM sleep. This suggests a zombielike state yet again, for it is in the DPC that such brain-powered things a logical thinking and the ability to plan ahead come from—our executive function, if you will. Qualities, I'm sure we'd all agree, zombies have in short supply, as if their DPC were in a state of perpetual deactivation.

Of course much, if not all, of these fascinating discoveries will have very little bearing on the waking lives of zombies unless one is a believer in plasticity. The case for plasticity says that there is a relationship between the sleeping life of the brain and the waking brain, that the brain grows, develops, and changes at night, particularly during REM sleep. Proponents of this idea point to the amount of sleep that newborns require as, in part, proof that plasticity is a true and vital element of brain development. As Hobson and Pace-Schott remind us,

"Brain activity *in utero* and in premature infants consists almost entirely of REM-sleeplike states."

The paper cites a handful of studies done on rats and even kittens that indicate a connection between brain development and deep sleep. What we can take from these studies and the theory of plasticity is this: if we start with the idea that zombies have a functional brain and that is why we must shoot, batter, or otherwise inflict harm to their heads to slow them down, then we can see how a zombie who rests for seven to nine quality hours a night will be even more terrifying than he already is. The parts of the brain that are already active in the ghoul—specifically the amygdala—will, it seems, develop even more, while the areas in which the ghoul is deficient—ye olde dorsolateral prefrontal cortex and all that jazz—wither yet more. Scary!

However you refer to it, one thing seems clear: when it comes to issues of sleep, no behavior seems quite as zombific as sleepwalking. Some people even call it sleep terrors. We've examined the zombific implications of what happens when one is either underslept or, on the other side of the spectrum, perfectly, deeply well rested. So far, so good—these notions make sense. What makes almost no sense at all is when a mostly average person, without actually waking, gets out of bed and into his car, drives many miles in the darkness, breaks into his in-laws', assaults his father-in-law, stabs his mother-in law to death, and remembers none of it because he was, essentially, asleep at the time. The technical term for such behavior is homicidal somnambulism. I prefer the more colloquial: holy shit!

According to Dr. Mark Mahowald of the Minnesota Regional Sleep Disorders Center, the states of wakefulness and sleep are not mutually exclusive. "It's really an admixture of wakefulness and nondream sleep" that causes one to sleepwalk, Dr. Mahowald told me in a phone conversation. "The parts of the brain that can perform complex behaviors are awake, the parts that are aware of what that behavior is and can form memories of it are asleep." Dr. Mahowald noted, as an example of what he means, that the medial frontal cortex, which helps us understand our behavior and remember what we've done, "clearly shuts off during sleep."

While it might seem remarkable that a sleepwalker could get out of bed, drive a car, and then kill someone, Dr. Mahowald suggests that we all have it in us. It turns out that sleepwalking behaviors are all "primitive or overlearned behaviors." Driving a car is an overlearned behavior, and during normal waking life, few of us even pay much attention to what we are doing when we are behind the wheel.

The fact that we can operate at all in a half-sleep state is due to complex neuronal activity derived from central pattern generators located in our nervous systems. These generators kick in when the brain is not capable of delivering the information we need to perform certain tasks. Dr. Mahowald said this is what is behind the strange phenomenon that has given rise to one of our most enduring metaphors, a chicken sans head running around. In those instances, the chicken's central pattern generators have instructed the poor bird to run.

Researchers at the Minnesota Regional Sleep Disorders Center have determined that sleepwalking and sleep disorders are not a psychological condition but rather a common part of the "human condition," as Dr. Mahowald puts it. He says about 5 percent of the U.S. population can be considered sleepwalkers. And about 2 percent of the adult population engages in violent behavior while in a half-wakeful daze. The episodes of violence originate with a disorder of arousal wherein the sleepwalker arises and enters a state known as automatism that is primarily a genetic function or brought on by sleep deprivation. Stresses such as the fact that you were just turned into a zombie can also trigger similar behaviors.

There have been some grisly cases involving homicidal somnambulism. Among them was one alluded to earlier, the 1987 murder by Kenneth Parks of his mother-in-law. Parks's lawyers pointed to an incredibly strange electroencephalography, or EEG, reading of Parks's brain's neuronal activity and a lack of motive in building his defense. It worked. Parks had committed a crime, but he had done it when the part of his brain that has awareness of actions was still asleep. He was acquitted of the crime. In 1997, Scott Falater was not so lucky. But then Falater had, after stabbing his wife forty-four times, stashed the knife with the clothes he wore during the killing in the trunk of his car before disposing of all of the evidence. His actions were considered to be beyond the simple or primitive behaviors a normal sleepwalker is capable of, and he was sentenced to life in jail for the crime of first-degree murder.

One thing Falater did that true sleepwalkers also do is attack a person near him. Dr. Mahowald says that the victims in these cases are almost always close to the perpetrator of the crime. The aggressor, he says, "almost never seeks out a victim." (There are, of course, the exceptions to this rule. Parks, for instance.) In this way, too, the behavior of a homicidal somnambulist resembles that of a zombie. For the undead are, if nothing else, equal opportunity flesh chewers. They do not seek out victims so much as bump into them in the parking lot or the men's room or a shopping mall, and, without giving it much thought at all, attack. It's as if the part of their brain that understands behavior and is capable of storing memories is asleep.

We established early on that zombies tend not to sleep, so what in the name of all that is holy does any of this have to do with zombies? Well, a lot, actually. So far we have established that the way in which human sleep behavior manifests itself can, at times, be most zombielike in nature. But we've saved perhaps the most important bit of information for last, and this, seriously, is stuff that could save a zombie's life. I am talking about fatal familial insomnia, which is totally not cool. It is, in fact, deadly. It can kill.

Fatal familial insomnia begins, as so many things do, in the section of the brain called the thalamus. For it is there that the brain and the body communicate, where the brain might, after a long night of Jäger navel shots and Mexican donkeys (not a euphemism), suggest that the body get a little shut-eye. For most of us, this is not a problem. But for those suffering from fatal familial insomnia it is a gigantic problem. Those sufferers have developed a nasty little prion that's blocked the thalamus from working properly. Soon, sleep becomes impossible, and within three years the sufferer dies, but often it can happen much faster than that.

Before death there are several less final but devastating stages of the disease to contend with. First comes almost half a year of debilitating panic attacks and the development of numerous new fears and phobias. This gives way to a period of hallucinations, which ensure that sleep will no longer be an option. It is around this point that the patient ages dramatically and can lose an incredible amount of weight and begin to resemble a zombie. Finally, during the last several months of the disease, a form of dementia descends and the inflicted one loses the ability to speak. Eventually the brain shuts down completely.

Zombies, consider yourselves warned—if you go on and continue your current lifestyle trends, you could be entering a world of pain and, maybe even worse, death. Luckily, doctors have only discovered fatal familial insomnia in roughly fifty families to date. But if your family is unlucky enough to suffer from the disease, you will have about a 50 percent chance of contracting it, too. One other neat thing about it: there is no cure. Just as with zombies.

So far in my life (knock on wood), I have not encountered a zombie invasion. Should it happen, I could see how at first it might seem as though I am only having a hallucination, a dream. I must be asleep, I might tell myself, because otherwise this thing that is very much human but not human, and intent on attacking me is real and that cannot be. Once I shook off this hazy notion and realized that I was actually awake and thus in pretty deep trouble, I would commence running. If I were fortunate enough to escape, I would immediately face another disturbing enemy: sleep. And I would not be alone.

In his chronicle of the battle against the beasts, *World War Z*, Max Brooks speaks with many folks who've faced down the zombie threat and seen their hours of rest evaporate. "I was exhausted, nervous, horribly sleep deprived," says but one of them. "I couldn't rest in the conventional sense." And so, one last cruel irony: just at the moment when we humans could most use a decent night of sawing logs, the very thing we need our strength to fight destroys our ability to sleep. And the entire wretched cycle starts anew.

In a related section of text that remains one of the most famous pieces of writing ever produced, a rather forlorn fellow announces:

> To sleep, perchance to dream. Ay, there's the rub,
> For in that sleep of death what dreams may come,
> When we have shuffled off this mortal coil,
> Must give us pause. There's the respect
> That makes calamity of so long life.

And it is thus that one of the world's all-time great insomniacs toils. There is little that can be done to make the Dane's lot less unsavory, and yet consider this: if he goes on not sleeping, as we've seen, the effects on his brain can be calamitous for sure. But if he does doze off, even for a minute or two, then he will be, like the rest of us, defenseless against a zombie attack.

EIGHT

BEE AFRAID, BEE VERY AFRAID

...

Examining real-life parasites that turn their victims into mindless zombies.

■ ■ ■ ■ ■

PARASITES—are there any creatures in the world that are more gruesomely insidious (other than zombies)? These burrowing, stinging, chewing, capillary-bursting, intestine-and-brain residing, microscopic beings are everywhere. Like *everywhere*. You thought the swarming hordes of *I Am Legend* or the vast, hungry armies of *World War Z* were bad? My poor, naive, soon-to-be infected friend. To borrow the poignant words of Bachman, Turner, and even Overdrive: you ain't seen nothing yet.

While zombies mostly thrive near gravestones or in little-used storage closets housing, inexplicably, the highly toxic cryogenically frozen remains of medical specimens, parasites happily find homes everywhere, from still water to caterpillars to even the poo of cats! See what I mean about the insidious part? But I fear I am giving the wrong impression about these little buggers. The truth is that they have far more in common with the undead than they have differences.

This phenomenon can be seen in the masterwork on the subject of parasites, *Parasite Rex*, by the man many people consider the godfather (or godorganism, if you prefer) of parasitic chronicling, Carl Zimmer. (Lest you doubt that Zimmer is one serious *macher* of minimunchers, you should know that he actually has a species of parasite named after him, the tapeworm *Acanthobothrium zimmeri*.) In dissecting parasites, Zimmer brushes up very closely to certain flesh-curdling characteristics of and insights about our friends the zombies.

A few resonant choice bits from *Parasite Rex*:

- In writing about a fellow parasite enthusiast and academic (Zimmer is a lecturer at Yale), he says, "He discovered that while parasites might be grotesque, they were also the most interesting creatures he had ever encountered."
- "Parasites live in a warped version of the outer world, a place with its own rules of navigation, of finding food and making a home."
- "Drinking blood is not easy. When a mosquito lands on your arm, it has to drive its proboscis through the tough outer layers of your skin and then snake it around for a while to find a blood vessel. The longer it takes, the better its chances of getting slapped and being reduced to a bloody smear."

Swap the words *zombie* and *the undead* for *parasite* and *mosquito* and I think you will agree that it all rings so very, very true. This is terrifying because pretty soon you are

imagining microscopic zombies lurking *everywhere*. And that's not that far from how the world really is. Zimmer and others have written extensively about a certain parasite we will explore in this chapter that lives inside roughly a third of us and in as many as 90 percent of Europeans in some parts of the world. And get this: it lives in our brains!

Let us examine the parasite that more than any other parasitic wasp—yes, every last one of the two hundred thousand species of such flying bandits—is and acts most like a zombie. This parasite is so much like a zombie that its victims are known as zombie cockroaches.

The creature in question is the *Ampulex compressa* species of wasp. Better known as jewel wasps, these serious stingers are popular in the more tropical areas of Africa, India, and on the Pacific Islands. Like Floridian zombies, moaning every morning for the early bird brain special, they thrive in warm weather. And also like zombies, the jewel wasps require the bodies of victims to help them perpetuate their species. Here's how they do it.

First the wasp finds itself an unsuspecting cockroach ambling down the boulevard. Sneaking up on the roach, the wasp jabs its stinger into the soft underbelly of the insect. This, as you might imagine, stuns the cockroach. Next the wasp makes a second attack on its victim, now plunging a stinger directly into the roach's brain and injecting its toxic zombifying solution straight into the skull. Within seconds the cockroach is paralyzed, unable to move and maybe even unable to think. "It's lost its will," Carl Zimmer explained to the National Public Radio affiliate WNYC in 2009. "It's a puppet. It's become a zombie, basically."

That's just the beginning of the wasp's wicked plan. Once the cockroach has been effectively zombified, its attacker grabs hold of its antennae and pulls it along back home, burrowing down into its subterranean lair with the cockroach in tow. "Like a dog on a leash," Zimmer notes, though I might amend that slightly and say it's like a zombie on a leash! Not nearly finished with the torture session, the jewel wasp then lays an egg on the cockroach's abdomen. By this point the roach is fully alive and functioning but cannot escape, as the wasp's injection has robbed it of its motor skills.

In 2007 Frederic Libersat of Ben-Gurion University in Israel studied how exactly the wasp's serum stops the cockroach in its tracks. What he deduced was that the nasty cocktail that's blasted into the cockroach's brain contains, among other things, a chemical that effectively disables, or blocks, certain neurotransmitters in the cockroach's brain. One of the transmitters that is most affected is called octopamine, which plays a big role in helping cockroaches perform complicated physical functions such as walking or running (or scurrying under the toaster as soon as the lights are flicked on). The Ben-Gurion researcher arrived at this conclusion after he observed the effects of injecting cockroaches with a chemical mixture that undid the blocking of the octopamine neurotransmitters. When that drug was administered, the previously zombific creature was once more able to walk. "I think the most likely explanation" of the phenomenon of zombie cockroaches, Libersat told *New Scientist* magazine, "is that a component of the toxin affects the expression of genes that regulate the activity of these neurons."

The *New Scientist* reporter working on the story then had the good sense to ask the question I think we are all wondering right about now. Namely, "Could octopamine become a possible antidote for future humans turned into zombies by, say, invading aliens?" Sadly, no, answered a German neuroscientist named Hans-Joachim Pflüger. Except he said it like this: "Our brain is of course much more complex, and we use different neurotransmitters. But new research shows tiny quantities of octopamine exist in the vertebrate spinal cord and do affect leg movement, so it will be interesting to see what exactly octopamine does in humans."

For now I'm afraid all we can do is marvel at the similarities between octopamine in cockroaches and the poisonous tetrodotoxin, the crazy stuff found in puffer fish that has been linked to the zombifying of humans primarily in parts of Haiti (and is covered in chapter 5). By fiddling with both formulas, it seems, would-be zombie slave owners can erase the fleeing instincts and abilities of their victims.

The jewel wasp takes it to a whole new level. Because, let's remember, once it has positioned its zombie cockroach in an underground prison cell, it proceeds to lay its eggs on the stomach of the roach. Once the wasp's larva hatches, it sure doesn't have to go far for dinner. Dinner, in fact, is just beneath it. That's right, the newborn wasps feast on the belly of the cockroach, chomping straight into its guts and then devouring it from the inside out over the next eight days. Amazingly, during that time the wasp young'uns eat their host in a manner that keeps them alive so that they won't be forced to feast on a rotting corpse. "Parasites are very careful," Zimmer notes. "They won't eat vital organs that would kill it."

It's around this point that the WNYC radio host with whom Zimmer is speaking declares, "That, to me, sounds like the purest description in nature of evil that I can imagine."

He's got a good point. I mean, in a way, the *Ampulex compressa* species of wasp has out-zombied zombies. At least when a zombie attacks you it all happens very quickly, without any of that stunning business and then the dragging into a dark, underground burrow to essentially torture you for eight days so that it can create more zombies. With a zombie, boom—you're a goner! One glaring difference here, of course, is that zombies will turn you into a zombie, just like them. In the case of the jewel wasp, it is only using a zombified critter to help it reproduce before leaving the zombie host for dead.

In *Parasite Rex* Zimmer concludes that "the most vicious and unneighborly behavior of all . . . can be found among some of the parasitic wasps that so impressed Darwin. This shouldn't come as too much of a surprise, given the gruesome way the wasps treat their hosts." Reminds me of the hot, but deadly, 406 that tried to slurp up poor Columbus in *Zombieland*. Talk about unneighborly!

It's hard to say, then, which is worse, zombies or *Ampulex compressa*. I suppose it comes down to personal preference: when you are feasted on by zombies, would you rather become one yourself or simply be left for dead? Is a future of so much wandering, moaning, and flesh-suckling too terrible to imagine? Before you answer, consider the circumstances surrounding the rather dastardly protozoan parasite known as *Toxoplasma gondii.*

As a 2003 paper from the medical journal *Emerging Infectious Diseases* put it, "*T. gondii* is of special interest because of its known affinity for brain tissue and its capacity for long-term infection starting early in life." Sound like someone we know? *Toxoplasma gondii* possesses other similarities to zombies, as well. Remember all that stuff earlier about the insidiousness of parasites? Well, *T. gondii*'s got that in spades. And also remember about the organism born out of the feces of cats? Check and mate.

When a feline takes a dump it also ejects a microscopic egglike thing called an oocyst containing the *T. gondii* parasite. These tiny sacks, unlike zombies, are just fine kicking it until an unsuspecting creature—a bird, perhaps, but often something of the ground, such as a rat—comes along and eats it. That's when the parasite reveals behavior that's decidedly undead. The oocysts burst open, freeing the protozoa, which immediately disperse inside the body of the new host. Each of the protozoa looks for, and eventually finds, a cell to call its own.

Let me turn it back over to Carl Zimmer here for his rather zombielike description of what happens once a protozoon locates a cell to its liking: "Once *Toxoplasma* has invaded a cell, it starts feeding and reproducing. After it has divided into 128 new copies, it tears the cell open, and the new parasites spill out, ready to invade fresh cells." This is similar to the scene in any number of zombie flicks when what seems like a manageable number of freely circulating zombies suddenly and quite horrifyingly becomes an overwhelming army of tendon mashers.

In some ways the *Toxoplasma* is even smarter and thus more sinister than the ghouls who would devour us. Because once they've completely invaded their host—just as we saw with the jewel wasp larvae that would not destroy the cockroaches until they'd gotten exactly what they needed out of them—the parasites are content to build tiny membrane shelters in which to live. The *T. gondii* organisms remain inside the attacked host until the animal is hunted and devoured by a cat, which completes the cycle of life for the parasite as it then sets about producing males and females and new oocysts to be eventually excreted by the cat.

All of this can take years to accomplish. It shows admirable patience and foresight on the part of this parasite—and could be a lesson to all of those unthinking, unplanning zombies looking to bite into the first host they can grab. But lord knows it isn't easy: the *T. gondii* must actually manipulate and alter its host's immune system for it to be healthy enough not to succumb to parasites. The *Toxoplasma* does this by riling up the mobs of aggressive, or "inflammatory" to use Zimmer's word, T cells that would like nothing more than to wipe out the *Toxoplasma*. And in fact they are able to destroy a number of them—but the parasites that are living in their cyst shelters are able to safely survive. This has the effect of keeping the *Toxoplasma*'s numbers down and subsequently keeping the tiny organism's host healthy. For, as Zimmer notes, "If the parasite were to multiply madly, grinding up every cell in the host's body, it would find itself inside a corpse rather than a living host. That would hardly be the sort of thing that a cat would want to hunt." In this formulation, the parasites are not so much zombies themselves as they are itty-bitty assistants to kitty-zombies. Sounds sort of cute, but it's terrifying.

Speaking of fear and behavioral manipulation, it's worth mentioning yet another way that *T. gondii* casts a zombie-like hold over its host. These next remarkable findings come courtesy of a 2007 study and paper done by the National Academy of Sciences (NAS). The paper's authors noted that "after an acute infection, the protozoan parasite *Toxoplasma gondii* latently persists in the *brain* for the life of an infected host, offering an opportunity to study behavioral manipulation hypothesis."

Given that opportunity, the NAS researchers wasted little time in uncovering a fascinating wrinkle to the *T. gondii* mystery. What they determined was that, in general, animals like rodents have an innate fear of and an aversion to the scent of cats, specifically to the urine of cats. This fear is so great that rats will do almost anything to avoid contact with cats. It's the ultimate survival mechanism and it's biologically motivated. The NAS scientists' report on the rodent brain shows us how they maintain a "neuroanatomical circuit comprising the medial hypothalamic zone and associated forebrain structures. These forebrain inputs correspond to those emanating from the ventral hippocampus and the septum on one hand and the medial and basolateral amygdala on the other."

The rats can't help themselves; they are hardwired to detest the very beasts that would prey on them. And there's that tireless amygdala again, which we all now know factors very prominently in any neurobiological discussion of zombies.

In conducting their research, the NAS team looked specifically at the way the *Toxoplasma* parasite manipulates this hardwired circuitry. What they discovered was that the parasite is able to interrupt the innate processing of fear of cats in the rats' brains. Moreover, they observed that rather than fearing the felines, the rodents were actually charmed by them. They developed an attraction to the kitty pee! This effectively helped the parasitic organism lure its host back to not only the source of the infection but to the animal that must consume the host in order to complete the parasite's life cycle. Talk about insidious?

Keep in mind that this does not mean that the acutely infected rats were then rendered devoid of all fears. Oh no, when it comes to behavioral manipulation, the parasite is far more skilled than that—it limited the fright inhibitions to the feline's pheromones. This suggests a level of intelligence—or whatever you call the brain power of minuscule organisms—that zombies can only dream of. Well, zombies other than *Day of the Dead*'s Bub who memorably appeared capable of learning complex behaviors.

While thoughts of *Toxoplasma gondii* might appear to drift a bit from a hard-core discussion of zombie characteristics, the important intersection of those two things is clearly this behavior manipulation. It does not require too great of an imaginative leap to think of zombies as parasites and their victims as hosts. In simple terms we can watch as the zombie bites its victim, wiggling its fangs into the flesh in a manner that suggests what a jewel wasp does to a cockroach. Once the ghoul's saliva explodes inside human veins—in a way that brings to mind the protozoan cysts—the behavioral manipulation ramps up to a speed that would make your average *Toxoplasma* jealous. And that is at the bleeding, scabrous heart of what a zombie hopes to do—contort the hell out of its host to the point that its host becomes just like it. At which point the life cycle (death cycle?) of the zombie begins again.

To see the National Academy of Science writers sum up their findings is to glimpse the corollaries between parasites and the undead: "The core of parasitism is the ability of an organism to exploit its host. According to the manipulation hypothesis, a parasite may be able to alter the behavior of its host for its own selective benefit. Such selective behavioral change is proposed to increase reproductive success of the parasite, usually by enhancing its transmission efficiency." That is to say, run! The parasites are coming!

Moving away from the animal world, let's look at the ways in which parasitic organisms—particularly *T. gondii*, fond as it is of brain tissue—can take a perfectly fine human being and turn him into something else entirely. Perhaps the most fascinating manifestation of this has to do with a study undertaken in 2003 by E. Fuller Torrey of the Stanley Medical Research Institute in Bethesda, Maryland, and Robert H. Yolken of the Johns Hopkins University Medical Center in Baltimore on the connection between the *T. gondii* parasite and instances of schizophrenia. What they discovered is quite a connection and not a tenuous one at that. Early in their report, the authors write:

> Some cases of acute toxoplasmosis in adults are associated with psychiatric symptoms such as delusions and hallucinations. A review of 114 cases of acquired toxoplasmosis noted that "psychiatric disturbances were very frequent" in 24 of the case-patients. Case reports describe a 22-year-old woman who exhibited paranoid and bizarre delusions ("she said she had no veins in her arms and legs"), disorganized speech and flattened affect.

First, that description of the twenty-two-year-old woman—I don't think I have to point out just how zombie-ish that sounds, right? Not to trivialize something as grim and unyielding as schizophrenia, but in the context of this book I think it is worth pointing out how, at least metaphorically, a schizophrenic's troubled state does sound a lot like the zombie experience. A schizophrenic is himself, but also not himself; the same can be said of a zombie. The brain of the schizophrenic is in many ways intact, but the wiring has been tangled; the same, again, we can see in the lives of many zombies. Schizophrenics often become obsessed with a particular task, action, or behavior in a way that seems out of whack to the nonschizophrenic; zombies are nothing if not obsessive in their pursuits.

While Torrey and Yolken are not yet dead certain of the connection between the parasite and the schizophrenic, they present plenty of compelling data to support the thesis. They note that in France the population is infected by *Toxoplasma* at a rate that far outstrips many other parts of the world due to a more laissez-faire approach to undercooked meats and unpasteurized milk, which can host the parasite. The French also suffer from schizophrenia at roughly a 50 percent higher rate than their neighbors in England. Ireland also has a high incidence of both *Toxoplasma* and schizophrenia. While it could be argued that such evidence of a link is merely anecdotal, the researchers point to hard science, too, in making their case. They write:

> Neuropathologically, studies of *T. gondii* in cell culture have shown that glial cells, especially astrocytes,

> are selectively affected. Postmortem studies of schizophrenic brains have also reported many glial abnormalities, including decreased numbers of astrocytes. Similarly, animal studies of *Toxoplasma* infections have demonstrated that this organism affects levels of dopamine, norepinephrine, and other neurotransmitters, which are well known to be affected in persons with schizophrenia.

Got that? No, me neither. Essentially, that's a lot of brain matter that helps control and support neurons and neurotransmitters. They help with the moods and the thinking. *T. gondii* and schizophrenia have similar detrimental impacts on these regions of the brain.

Torrey and Yolken are quick to point out that their study was far from definitive. They are aware that the correlations they draw in the paper could be accounted for in other ways. For example, they point to an additional study that reminds us that hospitalized schizophrenics often work or spend time in gardens on the grounds—a location that is also popular with neighborhood cats. This might mean that the schizophrenia was present in the patient before the patient encountered an environment that could trigger a case of *T. gondii*, and so the parasite did not play a part in the onset of schizophrenia.

More recently, however, the work of the two Maryland scholars was backed up by additional studies. The new research was conducted in 2009 by a team of lab-coat wearers operating out of the University of Leeds' Faculty of Biological Sciences. Building on the statistical suggestion that there is indeed a connection between toxoplasmosis and schizophrenia, the Leeds researchers determined that the parasite was capable of producing dopamine, which directly affects a person's mood, motivation, anxiety, and sleep patterns, among other behavioral aspects. The Leeds study found a possible link not just between the parasite and schizophrenia but also with other neurological disorders, such as Parkinson's disease and Tourette's syndrome.

Research in this field is ongoing, and it is extremely difficult to unravel the many mysteries of schizophrenia. But one thing so far seems sure: the parasite *Toxoplasma gondii* can and does change humans. It can be argued that it makes humans less in control of their faculties, less capable of independent action and thought, less able to exert all of the many wonderful motor skills that they possess. It makes them less likely to have a steady emotional life. It makes them less human.

So what are we ultimately talking about when we talk about parasites and zombies? To my mind, we are talking about brothers that are more than spiritual, that have physical attributes in common, as well. Both organisms require an additional being in order to survive. You might call that other body a host, a human, or a human host. Both organisms, upon encountering their hosts, alter them if not completely then certainly irrevocably. Parasites and zombies enter from the outside, but they change you, me—us—from the inside out.

Once you've encountered a zombie—or a parasite—it is safe to say you will never be the same again. Although I suppose you could say that most jam bands will change you, too, and not in a good way.

NINE

FEAR AND LOATHING IN ZOMBIE-TOWN

...

Mass hysteria, post-traumatic stress disorder, and the undead.

■ ■ ■ ■ ■

IT was just another Peruvian summery Saturday night in 2007. Until it wasn't. In the distant town of Carancas in the Puno region of the country, some 800 miles south of the capital city of Lima, there was a brilliant flash of light in the sky, followed quickly by a booming noise, and then an eerie, final silence. The kind of silence that can mean only one thing: trouble. Well, it can mean two things, actually: trouble and there might soon be a zombie outbreak that will threaten to turn into a full-blown epidemic, which will ensnare us all in its organ-squirting trap and render us dead-eyed monster things before Sunday brunch. That is what such a silence generally portends.

In the case of the remote Peruvian village, however, the flashing of light, the booming noise, and the sudden silence was not the first act of a zombie play for our parts but rather the early hours of what would come to be known as a curious meteor strike. The massive and incredibly hot rock smashed into this small slice of the world and left a gooey brown crater that measured 98 feet wide and 20 feet deep. It looked like Satan's swimming pool. It had tall crumbling muddy walls the color of mahogany, and sludgy liquid bubbling up looking like the world's least delicious bowl of melted chocolate. In other words, the hole was a must-see for anyone who lived in the area.

People came in droves to gawk at the hideous ditch. Things like this didn't happen every day, after all. The event soon gave way to extensive debate among experts about nearly every aspect of the meteor crash, including whether or not it was, in fact, a meteor at all. This proved at least a couple of things: (1) what really happened that Saturday night would be hard to determine, and (2) experts love to debate.

What was immediately clear, though, was that many of the Peruvians who wandered over to take a gander at the hot mess not long afterward began complaining of similar symptoms of illness. They experienced headaches, wicked sore throats, irritated eyes, nausea, and vomiting. By one count, as many as six hundred citizens of Carancas fell ill very soon after the meteor, or whatever it was, fell to Earth.

The experts had plenty of theories about why that was. The most popular explanation was that when the big ball of icky matter struck ground, it effectively unearthed rich deposits of arsenic, which, as if this were a sequel to *Return of the Living Dead II*, had contaminated the atmosphere with noxious fumes that sickened visitors to the site. Others wondered whether perhaps the rock itself hadn't emitted some sort of poison gas. Then there were those who were skeptical, disdainful, or pragmatic, depending on your point of

view. This group of experts suggested that there were no toxic fumes, no dispersed arsenic, no biological disease at all. These experts believed it was all in those six hundred heads. "Those who say they are affected are the product of a collective psychosis," Puno's chief of the health department said to the *Los Angeles Times*.

In an interview with the Web site Space.com, a social anthropologist named Benny Peiser from John Moores University in Liverpool, England, elaborated on this idea: "The Peruvian event seems to be a rare case where we may be witnessing collective anxiety that is approaching near hysteria." Peiser went on to say, "The majority of the affected [in Carancas] hinted that some of the mass anxiety is due to fear of imminent impacts and psychological stress, which is not surprising given the premature speculation and media hype."

It may have been a controversial opinion to arrive at—especially if you were one of the six hundred vomiting victims—but in Peiser's assessment, it seemed that we were finally close to making sense of this strange occurrence. His invocation of the media's role pointed to a basis for the outbreak of illness in the science surrounding cases of mass hysteria. Peiser elaborated on this position:

> In recent years, there have been numerous cases where alleged meteorite falls were linked to mysterious explosions on the ground—only to be proven wrong. One of the main reasons for the significant increase of such claims is almost certainly due to the growing media interest in the cosmic impact risk. It is part of human nature—and extremely tempting

for the news media—to hype any event that initially looks mysterious.

Put in simpler terms, a Peruvian who lived in the area near the crater spoke for his fellow villagers when he told the BBC, "We don't know what is going on at the moment, that is what we are worried about."

It is perhaps reflective of the strange, shape-shifting nature of mass hysteria that it is known by so many names. Take your pick: conversion disorder; hysterical neurosis, conversion type; mass psychogenic illness; mass sociogenic illness; or simply epidemic hysteria. Whatever you want to call it, mass hysteria has been around for a while—the first recorded detection dates from 1374—and can be brought on by occurrences from alleged water pollution to a Martian invasion to my personal favorite, "the phantom anesthetist of Matoon." As a 1997 article in the journal *Epidemiologic Reviews* (you know you subscribe to it) noted, "Too often, it is the media-created event to which people respond rather than the objective situation itself."

This notion certainly resonates with the experience of our housebound, would-be heroes of *Night of the Living Dead*. For when Ben and Barbra first arrive at the secluded farmhouse in which they seek refuge from the mysterious monsters, there is a current of calm running through them as they attempt to understand and deal with their fear. (True, in Barbra's case the calm is actually a full-blown episode of catatonic shock, but isn't that just a really intense kind of calm?) Ben, meanwhile, sets about systematically securing the house, piling furniture in front of doors, boarding up windows and then boarding them up again. The relatively under control scene is disrupted, of course, by a trio of survivors who soon wander up from the basement, where they had been hiding from the undead. The three new characters are a middle-aged couple, Harry and Helen, and a mild-mannered younger man, Tom, who has a nifty mod haircut.

Harry, we can see right away, is a hothead, and we watch as he "infects" the others in the house with his hysteria. Ben and Harry have differing ideas about the best way to proceed: Harry says they will be safer in the basement, behind the thick cellar door; Ben says that is a sure way to get trapped with no way out. Before long they are really going at it, amping up the anxiety in the room.

The situation finally comes to a boil when the five non-zombies are able to tune in to a television report on what the hell exactly is going on. The initial analysis from the newscaster is far from comforting: "It has been established that persons who have recently died have been returning to life and committing acts of murder. A widespread investigation of funeral homes, morgues, and hospitals has concluded that the unburied dead have been returning to life and seeking human victims. It's hard for us here to be reporting this to you, but it does seem to be a fact."

Later, looking for more information on the radio, the Farmhouse Five get even worse news from the broadcast: "Civil defense officials in Cumberland have told newsmen that murder victims show evidence of having been partially devoured by their murderers. Consistent reports from witnesses to the effect that people who acted as if they were in a kind of trance were killing and eating their victims prompted authorities to examine the bodies of some of the victims."

As you know, what they found upon closer examination wasn't pretty. The reports come faster and more breathlessly by the minute, the media outlets delivering them at once aghast at and animated by the proceedings. The effects on those tuning in to the reports are clear: their fear is ratcheted up. The more they know, the worse it gets.

At one point a scientist walking to his car is hounded by a television reporter who demands an answer to the inchoate madness. The best the beleaguered chap can do is mumble something about how it all might be related to a satellite accident, which presumably may have infected the wandering ghouls. (Curiously, according to the Russian newspaper *Pravda*, Russian military analysts suggested that the Peruvian meteorite discussed in this chapter was actually a fallen U.S. spy satellite. They claimed it was the KH-13 aimed at Iran and that it was "destroyed in its orbit." When it crashed to Earth, the radioactive isotope used to power the thing, Pu-238, was intact, and that was the agent that had poisoned the Peruvians.)

The *Night of the Living Dead* scientist's televised impromptu press conference does little, of course, to quell the fears of viewers. If anything, it has the opposite effect. This is not uncommon, as the *Epidemiologic Reviews* article of 1997 makes plain:

> The stressful nature of the emergency response to the outbreak can enhance the problem. The presence of ambulances, fire trucks, television cameras, and workers in protective clothing can all add to the anxiety. Such activities confirm individual suspicions that the situation is dangerous. The appearance of researchers

> and other official visitors making inquiries after the fact can reintroduce the disease agent, anxiety.

Protective suits no doubt have a terrifying presence, as anyone who has seen *E.T.* can attest to. More recently, the great anthrax scare of the dark fall of 2001 reminded us all about just how anxiety-inducing protective suits can be. Not long after the anthrax reports hit the airwaves, researchers Robert E. Bartholomew and Simon Wessely noted the spike in anthrax-related claims in the *British Journal of Psychiatry.*

> Over 2,300 anthrax false alarms occurred the first two weeks of October, 2001. . . . In one case, a teacher and student reported minor forearm "chemical burns" after opening a letter and discerning a powder in the air. Subsequent analysis revealed no foreign substance in the envelope. There is a danger of responding to every incident in space suits and inadvertently amplifying psychological responses.

The differences between the kinds of scares we've talked about here so far—meteorites, anthrax, researchers—and a zombie attack should be fairly clear. There is less ambiguity surrounding the latter. That is, we generally know quite certainly when we have been consumed like Purina Human Chow. It's a point that may seem obvious to us now, but it sure wasn't obvious to the cast of *NOTLD*. Remember, at one point the TV newscaster felt it was necessary to tell listeners, "Reports, incredible as they may seem, are not the results of mass hysteria."

While in *NOTLD* the terror at least appeared to be very much inspired by the real threat of having a zombie devour one's innards, in many other instances mass hysteria can be prompted by happenings that are very likely not real. Take the case of a 1973 outbreak of hysteria in a Malaysian television factory as detailed in a paper published in the *Singapore Medical Journal*. All of the victims in this event, possibly brought on by a group hallucination, were young women between the ages of sixteen and thirty-one. In the article, the author L. P. Kok says that "later it spread to two neighboring factories. In the first factory there were a total number of 84 girls affected, 34 of them having 'attacks,' including screaming, shouting, struggling and in four cases, 'trance states.'" Almost half of the twenty-five women who agreed to speak with Kok for the article said they believed the outbreak was brought on by "spirit possession;" the others had no clue what could have caused it. The experience of a seventeen-year-old girl identified as Miss R. in the report is indicative of what many of the victims went through. She reported to work at the factory in the evening as she always did. Then, sometime around seven p.m. Miss R.

> saw a shadow in front of her . . . silhouetted against the wall. This shadow started to laugh in a low gruff frightening male tone. R. felt terrified, tense and confused and started to run and shout for help. She fell many times, but picked herself up and continued running. All this while she heard the sound of laughter and felt someone chasing her. She was brought to the nurses' room, but was unaware of what was happening. After

> some time she was sent home. On reaching home she did not recognize anyone. She screamed and laughed continuously. Her father burned incense and read a passage of the Koran to her, after which she went "soft." Half-an-hour later she regained consciousness and was herself again.

Switch a few details in Miss R.'s story and you end up with a picture that looks very much like what the catatonic Barbra went through out in that farmhouse in *Night of the Living Dead*. Not that the zombie attack was all inside Barbra's head. I'm just sayin'.

While issues having to do with mass hysteria are decidedly murky and difficult to treat, things clear up in a postattack world when dealing with hysteria's more knowable cousin, trauma. Post-traumatic stress disorder (PTSD) is a verified clinical condition involving overstimulated stress hormones. It is most commonly associated with soldiers who have fought in wars, but any type of traumatic event can trigger the disorder, so it is worth considering here. A 2008 paper in the *Journal of Psychiatric Research* described PTSD as the outcome when stress hormones in the body "overly strengthen the consolidation of conditioned fear, which is later manifest in durable fear responses to reminders of the event." In other words, just by later thinking about the traumatic event, one can spiral back into the physiological experience of the event. It can be a debilitating illness that once it has its grips on you is loath to let go. Happily for PTSD sufferers, there is new hope for treatment. And, as you might guess, that new hope is coming from the world of science! Yay, science!

The authors of the *Journal of Psychiatric Research* article determined that the aforementioned stress hormones, also known as beta adrenergics, that kick into overdrive in a PTSD patient can be blocked by injecting the drug propranolol, a popular hypertension medicine, into the bloodstream. The trick is that in order to successfully block the hormonal activity and thus reduce the effects of PTSD, the propranolol must be administered within six hours of the triggering traumatic event (read: zombie attack). But it is not possible to diagnose PTSD in that short amount of time—in fact, it takes about a month after the trauma has occurred for such a diagnosis to be possible.

So what do you do? The journal authors had an idea, wouldn't you know. It involved having their patients revisit the traumatizing event and retell the story so that they might physiologically reenter the situation, at which point researchers would apply the proper amount of propranolol. They tested the idea on nineteen subjects, administering 40 milligrams of propranolol as soon as the traumatic event was revisited and another 60 milligrams two hours later. To some patients, meanwhile, all that was given was a placebo. Aha—the old placebo trick! And guess what?

> A week later, in the psychophysiology laboratory, the subjects listened to their personal traumatic scripts and imagined the event while physiologic responses were measured. We hypothesized that subjects who received post-reactivation propranolol a week earlier would show smaller physiologic responses than those who received placebo.

In the case of one fellow whom two of the paper's authors studied, he had suffered PTSD after finding himself in the midst of a bank robbery and being pistol-whipped. The man had been terrified by the event, so much so that according to the *Wall Street Journal* he "gave up his long-time hobby of bird watching and broke up with his romantic partner. And he became house bound, because he felt unsafe whenever he went outside." The man, desperate to find a cure, underwent half a dozen propranolol treatments with the two doctors. After his fifth visit, "the man said he felt remote when reading the script he had written . . . two years after the treatment, the man has resumed his normal activities and says that although he remembers the events at the bank, he doesn't feel scared anymore."

I think we can all see how such a treatment would be helpful for victims who have survived zombie attacks. The man in the middle of the bank holdup said that that day he feared for his life—certainly the same could be said of many of those who were interviewed following the incredibly traumatic events of *World War Z*. I imagine they all would have benefited greatly from a few sessions under the propranolol needle.

Finding anything remotely like a cure for mass hysteria is a taller order, indeed. The available science suggests that while PTSD is a physiological condition, mass hysteria is more of a psychological one. But both conditions have at least one important factor in common: fear. Naturally, that skittish emotion is commonly generated in the region of the brain most associated with the zombie experience—that's right, our old friend the amygdala.

A couple of years ago researchers at the Ponce School of Medicine in Puerto Rico considered the expression of fear in lab rats. But rather than looking into ways to modulate the neuronal activity in the animals' tiny amygdalas, the scientists attempted to bolster the part of the brain that instructs such emotions as security and safety; in essence, the opposite of fear. Those emotions derive from the prefrontal cortex. The Ponce School professors found that stimulating the prefrontal cortex in the rats had the effect of minimizing their fear and anxiety. Just as the administration of propranolol on PTSD victims did not erase the past experience, but only helped inhibit the response to the trauma, the prefrontal cortex hot-wiring did not wipe out all fears in the rats but instead counteracted the fear-mongering going on in the amygdala.

That treatment may prove to be the best we can hope for when it comes to helping zombie-attack survivors. They will always have the fear—and, really, who can blame them? But maybe that fear can be shrunken to a manageable size.

Franklin Delano Roosevelt, during his Inaugural Address in 1933, famously said that the only thing to fear in this topsy-turvy world is fear itself (I'm improvising a bit here). That's a profound idea, beautiful in its simplicity. Too bad it isn't true. In addition to fearing fear we must also fear spiders, undercooked beef, and excessive mold stuck in grout. We must fear tornadoes, hurricanes, and the sun; armed militias, bandits, and cyclopes; contracting rabies, lockjaw, and the clap. Splinters, too, if they become infected, which they often do. Strangers who stand extremely close to you while talking loudly can be scary. So can squirrels. Yes, they can!

Most of all we must fear zombies because zombies are truly terrifying. They want to eat our faces off, which is really unnerving. So there is nothing to fear except fear itself, all this stuff I mentioned (and many other things I didn't mention), and especially zombies. The bad news is, once one of us fears zombies, chances are that fear will spread like an irrational disease, riding high on the airwaves of so many media outlets scrambling for coverage. There's not much we can do about that.

The good news is that if you somehow give zombies the slip or are able to make a dash for it and come out intact on the other side of the war, then it appears we now have a drug for you to help block the stress hormones that would have you living the rest of your days in a catatonic state. Now, I've never been to that state so I can't say for sure, but my hunch is that's not a place where you want to live. Or even visit on a long bender of a weekend while drinking too much and sleeping with too many strangers. Although that may not be half bad.

Mad Limbs #1

■ ■ ■

Best if read by a grizzled old fella. A gravel-voiced man of deep wrinkles and deeper traumas. Holding a gun. By a fireplace. At night.

Old Salt

My ____________________ (MALE FAMILY MEMBER) heard 'em first. Heard 'em moving in the dark. Said it sounded like a(n) ____________________ (ANIMAL) molesting ____________ (NUMBER) ____________________ (PLURAL ANIMAL). We didn't listen. He was always saying crazy things like that.

Then we heard him screaming. Wailing "______________ (EXPLETIVE)!" in the night. Over and over. Howls in the night, before one of those ____________________ (ADJECTIVE) undead bastards ripped his ________________ (BODY PART) off. I can still hear it sometimes. Every time I ________________ (VERB, PRESENT TENSE).

Then things got bad.

It was just ____________________ (FEMALE FAMILY MEMBER) and me in the house. They went for her first. And she looked at me. I remember her

face. Like a ____________ ____________
ADJECTIVE SMALL PET

crying in the rain.

I fought 'em with everything I had. I'm yelling, "Come and get me, you goddamned ____________ ! You cannibal
PLURAL NOUN

goddamned ____________ !" Wasn't any use. You'd
PLURAL NOUN

have thought it was ____________ the way they ate
CHINESE FOOD DISH

her brain.

And before I had a second to mourn her proper, I'm dragging her body into the ____________. The zombies
LOCATION

bust in; they're grabbing her, biting at her ankles. I drag her into the closet. The zombies follow. They're clawing at us, all hungry-like. I feel their ____________s on my
BODY PART

____________.
BODY PART

I'm not even thinking now. I'm a caveman now. I reach for the ____________ I've been keeping in there, I grip what-
WEAPON

ever I can with my other hand, and I just start swinging. Wasn't until later I realized I'd picked up my ____________. You
BODY PART

ever fight a horde of zombies with a ____________ and
SAME BODY PART

your dead ____________s ____________?
SAME FEMALE FAMILY MEMBER BODY PART

Believe me, it makes an impression.

So yeah, I've seen some things. ________________ things
ADJECTIVE

I can't never forget. And I'm staying here until they come back

for me. And when they do, I'll ____________________ with a
VERB

____________________ and say ____________________ with
NOUN EXCLAMATION

all my goddamned might, "You ____________________ can
DEROGATORY NOUN

stay until morning, but you'd best be gone by afternoon."

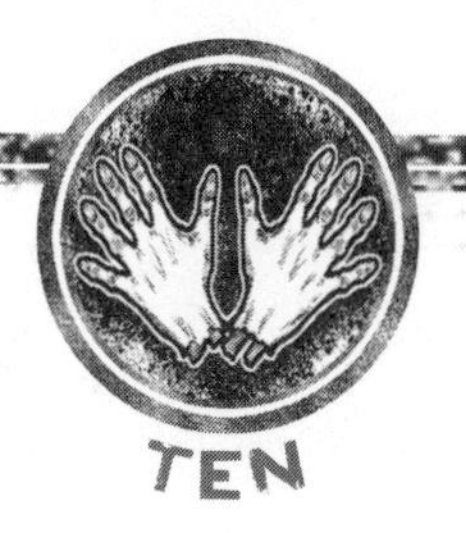

TEN

WAR OF THE WEIRDS

■ ■ ■

How do the undead hunt? How do humans flee? What being attacked by a rabid dog and living life like a slave-making ant can teach us about surviving a zombie outbreak.

■ ■ ■ ■ ■

NOW we must talk about cardio health. We must think about cardio fitness as it relates to escaping from zombies on the march, because make no mistake, they are in fine form themselves. If you plan to make a break and escape from the undead (no matter how quickly or how slowly they are pursuing you), you should know a thing or two about the hunters and the hunted, about those who are pursuing their game and those—like you and me—who are evading attack.

The history of hunting is long and bloody. According to the *Oxford Encyclopedia of Economic History*, it is approximately two million years old. It was around that time that "hominins show marked gut reduction and brain expansion relative to chimpanzees . . . and at the same time animal remains become much more common in the hominin archeological sites."

These physiological changes can pinpoint the start date of hunting, as the reduced waistline was caused by a more nutritious, low-fiber diet, and the surge in brain power was brought on by the fatty acids associated with the meat of animals. Eerily, it was at the dawn of the hunting epoch that our brains grew larger, thus providing an even larger bull's-eye for zombies on the prowl.

Taking note of the slimming torso and bulging brains, the *Oxford Encyclopedia* concludes: "The coincidence of these traits in time seems to strongly indicate an increased reliance on meat rather than plant foods in the diet of our ancestors during the period of early Homo." Notions of how to corner and kill prey have been with us for quite some time.

In general, however, humans have been on the pursuing side of the ledger, or, when sitting out a round, we have been happy to check out a little animal-on-animal hunting action. The rarer phenomenon, historically, has been instances of human-on-human pursuit, and rarer still, at least until this point, has been the undead versus the living.

This is somewhat new ground we are covering. But it is certainly not without precedence.

Take, as but one example, the early scenes of *28 Weeks Later*, the chilling sequel to *28 Days Later*. For it is there that we first encounter a small band of uncontaminated humans hiding out in what appears to be the lush English country-side. Interestingly, the handful of survivors have chosen to set up shop in an old rustic farmhouse, which indicates that they likely have never seen that ne plus ultra of zombie films, *Night of the Living Dead*. Had they seen the classic flick, they surely would not have chosen to bunk down in a shack

that so closely resembles the escape pad from the Romero masterpiece. In any case, the *28 Weeks Later* gang did just that and they would live (or, really, die) to regret it.

Among the farmhouse dwellers is a youngish couple named Don and Alice. They seem very much in love and even try to steal a quick kiss in the kitchen before being interrupted by dear ol' granny, who wanders in at exactly the wrong moment. But they are not so much in love that when the zombies come, Don determines the best route for evasion, leaving Alice to get mauled in the attic window.

The trouble begins, as it often does, with a child. Specifically, a small blond boy who bangs on the farmhouse door until Alice mercifully lets him in. The only problem is that the boy has been trailed, and soon enough the undead come knocking, too—except when they come knocking they tend to tear right through the walls.

It is at this point that a viewer could swear up and down that Don had previously read the paper "Co-evolution of Pursuit and Evasion II: Simulation Methods and Results" by researchers with the School of Cognitive and Computing Sciences at the University of Sussex and from the Center for Adaptive Behavior and Cognition at the Max Planck Institute for Psychological Research in Germany.

Much of that paper is dedicated to models and charts that on the surface have very little bearing on a discussion about a zombie's pursuit of a human prey, but there are some relevant passages. The authors, Dave Cliff and Geoffrey F. Miller, indicate that the act of pursuing can have very real physiological effects. Specifically, "eyes and brains can . . . co-evolve within each simulated species—for example, pursuers usually evolved eyes on the front of their bodies (like cheetahs), while evaders usually evolved eyes pointing sideways or even backwards (like gazelles)." While there can be little question that zombies don't generally resemble cheetahs, they do, like other humans, have eyes on the front of their bodies. Did these evolve in order to hunt other humans?

Don from *28 Weeks Later*, meanwhile, most certainly does not have eyes pointing sideways or even backward like a gazelle, but when sprinting across the expansive verdant meadow, which effectively serves as the farmhouse's front yard, with a team of cornerback-quick zombies in hot pursuit, he resembles no other animal as much as a gazelle. And luckily for him, Don turns out to be like a gazelle that can operate a motorboat, and thus narrowly escapes the closing-in creatures.

In a graph for their paper on simulation methods, Cliff and Miller arrive at a visual rendering of how Don managed to escape. On the *x* axis, they chart relative maneuverability, or turning ability. On the *y* axis, they plot relative speed, or acceleration ability. Cliff and Miller then deduce:

> The further into the upper-right zone [of the graph], the greater the chances of pursuers winning all contests (they catch up with the evaders, and out-turn them too); the further to the lower-left, the more likely the evaders are to do very well (the pursuers can neither catch them nor out-turn them).

So far so good.

The fascinating thing about this, though, is what comes later in the paper. It's there that Cliff and Miller contend that all things being equal—speed, maneuverability, and so forth—the evader can forever outpace the pursuer, assuming the evader had some kind of lead to begin with. Or, as the scientists put it:

> The pursuer and evader have the same accelerations and maximum speeds, which gives a clear advantage to the evader. This is because, as long as the trial starts with sufficient distance between the pursuer and evader for the evader to turn to face away from the pursuer before the pursuer can hit the evader, the evader need only perform such a turn and then accelerate to top speed to avoid being caught—the top speed is the same as the pursuer's.

This, essentially, is the principle that is employed when Don makes his break from the surprisingly nimble zombies in *28 Weeks Later*'s first several minutes.

Of course, what is not accounted for in this model is what happens when Don runs out of fuel, when he is too pooped to go on. As far as I can tell, the undead don't have the same problems of stamina that we humans encounter all too often. In Don's case, he dealt with this problem by hopping in the aforementioned motorboat to make his last-gasp getaway. This behavior, too, is covered in the same paper on pursuit and evasion. There, Cliff and Miller suggest:

> In many species of prey (i.e., evader) animals, there are often three distinct phases to an encounter with a predator (i.e., pursuer) animal. At first the prey will

be vigilant, conserving energy. When the predator comes closer, the prey will engage in linear fleeing, where it moves in a straight line at high speed, away from the predator. Once the predator is within some nearby distance threshold, the prey will then switch to the third phase, involving protean jinking and dodging behavior.

In Don's case, the protean jinking and dodging behavior involved an outboard engine, but it's different for different species, as the saying goes.

We can't all be lucky enough to run right into a waiting speedboat, of course, so we shoud examine some other options, should you find yourself needing to evade pursuing zombies. Let's look first at how one should generally deal with attacks by other wild animals, and perhaps in these examples, we can learn something about evading the undead.

Bears. Not all bears attack equally. Grizzlies, to begin with, are far more likely than black bears to want to eat you. Should you wander upon a grizzly in the wild, you are advised to keep your distance and hope the hungry, furry fella doesn't notice you. When that fails, you should attempt to create some space between you and the animal, but you must try to do this without running as that will indicate to the bear that you are prey that ought to be pursued. Speed walking is a good idea, no matter how silly it almost always looks. If you are holding a ham sandwich—or any food at all—discard it and pray that the bear is more interested in your lunch than in you as lunch. If the pursuit continues, stop and face the bear and throw your hands in the air like you just don't care (even though we know for a fact you do care)—this is meant to make you appear larger and ballsier than you really are and may actually scare the bear off. If that fails, you are actually advised to fall to the ground and cover your head and face. Or climb a tree. Or hop in a motorboat. All of these strategies should have the effect of diminishing the bear's interest. But while these ideas may save your life when facing a grizzly attack, the same sadly cannot be said about a zombie beat-down. The zombie wants you, not your ham sandwich, and since it eats less fish and berries and more brains than a bear does, it is not likely to lose interest in you, even if you're up a tall tree. All of which is to say, we must look elsewhere in nature for zombie evasion techniques.

Mountain lions. Your response to a mountain lion attack is much closer to what you will want to do to zombies than the above bear example is. And you will have plenty of chances to practice, as mountain lions have become seemingly more brazen than ever. How does one contend with this terror? One reliable Web site says to "use every weapon in your personal arsenal to defend yourself. Sticks, clubs, large rocks, mace, pepper spray, and weapons all need to be employed against a mountain lion attack." Boy, that sounds familiar. Why is such vengeance required, you ask? Because "once an attack begins, nothing short of killing or grossly injuring this animal is likely to stop it. Mountain lions do consider human beings potential sources of food and will generally prey upon smaller, less defensive humans." That's why. We may have found nature's zombie match in mountain lions, but let's examine one other kind of attack just to be sure we haven't missed anything.

Wild dogs. When confronted with this snarling menace, you should treat it is as you would a bear, only with more talking. Walk slowly away from the animal while keeping an eye on it and, in no uncertain terms, telling it to take a hike. If the pissed-off pooch still doesn't get the message, you may commence bashing it about the face with a stick. If that does not have the desired effect, try kicking it in the face, too. Actually, in the event of pretty much any type of attack, kicking the attacker in the face is generally a good idea.

By looking at these three examples of attacks that can occur in the wild, and borrowing from each, what emerges is a comprehensive plan for dealing with a zombie attack, especially when there is no motorboat handy. First, try to lie low and hope that the zombie/bear/mountain lion/wild dog does not notice you. Next, discard any food that the zombie/bear/mountain lion/wild dog might want instead of you—here, I am thinking of a brain omelet. Should you need further evasion techniques, now is when you should move as swiftly away from the zombie/bear/mountain lion/wild dog as you can without upsetting the creature. Try to intimidate the attacker by making yourself appear bigger than you are or by speaking in a calm but stern voice. Then climb a tree or begin to bludgeon the beast about the face, especially the head (more on this in chapter 12).

While these zombie evasion tips are right on the money, they also rely too much on the strength and daring of the average human, which is a dangerous proposition, indeed. To truly outsmart and outmaneuver the parading ghouls, we will need to first and foremost outthink them. And for that we must once again rely on science to bail us out.

Science totally has our back thanks to Italian genius Davide Cassi, who in 2009 published an equation-heavy article in the journal *Physical Review E* detailing the best way and the best place to avoid being turned into so much zombie stew. Well, he didn't really come out and say that was what he was doing, but when you title your paper "Target Annihilation by Diffusing Particles in Inhomogeneous Geometries," it's pretty obvious what you are up to, am I right? Cassi says right off the bat that "the survival probability of immobile targets annihilated by a population of random walkers on inhomogeneous discrete structures, such as disordered solids, glasses, fractals, polymer networks, and gels is analytically investigated." You don't have to have an advanced degree in spatial geometry to know what Cassi means. Random walkers? He means zombies!

In order to determine the smartest way to elude these randomly walking zombies, Cassi notes that we are dealing with a popular sort of reaction, specifically one that looks like the $A + B \rightarrow B$ process, where A is the immobile target and the reaction is used to indicate several possible variables in the hunt, including the physical environment and any possible chemical or biochemical events that may influence the outcome of the pursuit. What he means is that he is going to calculate which physical environment allows for a greater chance of survival among the pursued (humans) as they are being hunted (by zombies).

As you can imagine, to calculate this in any real and exhaustive way requires Cassi to scribble numerous charts, graphs, and formulas to arrive at a meaningful conclusion. But arrive he does. What Cassi discovered is that when random particles—or in this case, zombies—are moving through physical environments toward an unmoving target, the target will survive for a longer period of time if the environment in which the target is positioned has a more complicated topography.

This makes sense: if one is hiding, say, on the 50-yard line of a football field, a randomly mobilized entity—even a not-too-swift entity—will find or at least bump into the hidden target at a faster rate than if the target/human is hiding in an architecturally complicated environment, with a lot of walls to bump into and alleys in which to get lost.

This brings to mind the survival instincts and success of most of the fleeing humans in the 1978 version of *Dawn of the Dead.* The small but heavily armed crew famously make for the nearest shopping mall, a place with more than its share of complicated architecture that can act as blockades or deterrents to the random walkers: there are fountains, escalators, and twisting passages in the mall that the evading humans utilize to extend their stay. That proves manageable, particularly when the humans are being pursued by what a scientist in the film describes as "nothing but pure, motorized instinct."

Compare this to a wide-open football field like the one seen in the first few minutes of *Zombieland*, when an evader is easily run down, and you begin to understand the benefits of Cassi's work. At the end of his article, Cassi writes: "The detailed investigation of these aspects is fundamental to design optimal reaction strategies based on geometry, as well as to understand target decay processes in complex biological systems." In other words, Cassi means to show us the best place, in terms of the physical geometry and structural sturdiness of the space, to hide.

Could another approach to evading the undead be not to evade them at all? Here I'm thinking of the part in *Shaun of the Dead* when Shaun and the gang figure out that if you walk like a zombie (slowly, stiffly) and talk like a zombie (moaningly), then many zombies might just think you are a zombie. They use this technique most successfully and at one point safely move through the swarming streets of London and into the far more secure—and structurally complex—location of an unoccupied pub. This suggests a couple of things:

1. Zombies may not be that smart.
2. Perhaps we've been thinking about this whole escaping thing in totally the wrong way. Maybe instead of fleeing the fight, we should take the fight to the zombies.

Consider, for a moment, the slave-making ant, better known in some circles as *Polyergus breviceps*, for inspiration. At first glance the so-called slave-making ants may seem a sorry lot. That's because they have pathetically forgotten how to feed and otherwise care for themselves and their ant offspring. Instead, they've turned to other ant species for help—specifically, they find and prey upon colonies of *Formica gnava* and *Formica occulta*. Once they've located a colony suitable for attack, they enter and release pheromones and formic acids that have a calming effect on the *Formica* ants. (Some experts also believe the *Formica* ants mistake the *Polyergus breviceps* for themselves.) Sufficiently subdued, the *Formica* ants then can only sit idly as the *P. breviceps* steal their tiny ant eggs, or pupae, and bring them back to their own colony. When the eggs hatch and the new ants emerge, the *P. breviceps* effectively enslave the *Formica* ants and force them to forage for food for both adult and baby slave-making ants. It's a neat trick and one that would not be easy to replicate while battling zombies. But the principle is close enough to what Shaun came up with to suggest an alternative approach to simply fleeing.

Running and hiding will only get you so far. In both of the mathematical models examined here, the formulas suggested a measure of hope—the simulation devised by Cliff and Miller noted that if all things were equal, an evader could win over a pursuer, while Cassi observed the advantages of retreating to a geographically complex structure.

But in both instances, the hope only went so far. In the simulated pursuit, the fact that stamina and exhaustion were not initially factored into the proceedings suggests that the evader would not win for long. And while Cassi predicts that "target annihilation" would take longer in a physically varied structure, annihilation would ultimately still take place.

So where does all this leave us in our efforts to avoid being hunted by zombies like so many wayward hikers encountering hungry mountain lions? Mostly it leaves us in the same spot we've been in: doomed. Sorry to have to tell you that.

However, we can also take heart in knowing that there are options. We can look to Don in *28 Weeks Later* and see that despite the effects of exhaustion and limited stamina, there is a way to escape if you have a head start and can run as fast as fast-running zombies (and have easy access to a speedboat). And we can look to the *Dawn of the Dead* gang who when faced with a hideous outbreak of arm-munching ghouls, smartly went shopping. And, also, we have Shaun who, unwittingly taking a page out of the *Polyergus breviceps* playbook, more or less invaded a zombie colony, fooled its inhabitants into thinking he was one of them, and managed to stay alive in the process.

So, yes, random walkers can be terrifyingly unyielding (and far too interested in spleen sandwiches), but the finality of their convictions can at least be temporarily put off. Maybe buying a few extra minutes of time will be enough to allow us to come up with a serum so powerful it will wipe out any zombie that attacks. Or, short of that, we've always got shotguns. And, as it turns out, the physics of guns and ballistics is a subject we will explore at the end of the book. Lucky for us!

ELEVEN

ATTACK OF THE MUTANT ZOMBIES!

What can we learn from ghouls about mutation and radiation?

■ ■ ■ ■ ■

THERE are plenty of ways to be turned into a zombie. We've touched on a few already, but let's take a quick refresher course before charging ahead to explore one of the most scarily plausible zombification scenarios around.

First we have what is widely called the voodoo zombie type. This, you might recall, involves mysterious evil priests or witch doctors—mostly living in Haiti—who are fond of kidnapping living humans and injecting them with a wicked compound containing the poisonous toxin found in puffer fish, tetrodotoxin. This, ultimately, has the effect of disorienting the victims to the point that they become mindless zombie slaves. Effective but labor intensive.

A variation on this idea can be found in films and books such as *I Am Legend* or *28 Days Later*. In those productions, instead of a bad agent acting alone to enslave his victims, we have shadowy global government plots to infect the greater population with a virus, again by way of a highly potent bitch's brew. There are also menacing slug-worms from outer space as in *Night of the Creeps*—a less immediately worrisome issue in many ways.

Another way that zombifying can occur may sound less sinister, but rest assured it is no less insidious. This type of zombie is created when a person is forced to work the same boring job, day after mind-numbing day. We probably all know someone who was a pretty decent human being until he was hired to work at a really dull job, forced every day to rise early in the morning only to face a soul-stabbing commute on a packed and putrid subway car. His office is probably in midtown or maybe in the financial district. There are no doubt cubicles involved and a communal kitchen where all of the biggest sad sacks at the company eat their pathetic meals. These lonely suckers would cry daily into their ramen if the airless building they work in hadn't already sapped all the moisture from their tear ducts. These same eyes are ringed with dark circles, have sunk deep into skulls, and have the low throb of death in them.

But arguably the most popular circumstances for igniting an undead outbreak involves some form of radiation. Since at least 1968, zombie watchers have been certain that a horrific toxic spill would leave us deformed, haunted, and unable to form proper sentences. That's because in *Night of the Living Dead* the sudden and terrible undead onslaught is attributed to a radiation leak from a downed satellite. And it's also because events such as Hiroshima, the cold war, and the disaster at Chernobyl remind us of the unthinkably brutal, transforming capabilities of radioactive materials.

When the true zombie plague comes calling, history suggests it will on some level be due to radiation's ferocious streak. Perhaps we should get to know this ghoulish stuff a bit better.

Radiation is what we call the result of a form of energy moving from one source to another. Radiation can cause variations in the genetic code. If the variation leads to a viable organism, it is called a mutation. Radiation is all around us. The sun's heat radiating from a hot core reaches you and turns your forearms brown and foxy, as if you just came back from spring break. That's radiation. A microwave oven magically converts your breakfast burrito from a frozen slab of meat into a delicious meal. Radiation!

The trouble occurs when radiation leads to particles being mutated or ionized. This happens when electrons are refigured in their atomic homes (and there are lots of chances for this refiguring to occur—one uranium atom contains ninety-two protons and ninety-two electrons). When this happens, cells can undergo mutations, which can lead to serious and even fatal diseases, particularly cancer.

A few of the more common radioactive isotopes include uranium 235, plutonium 239, and iodine 131. These materials have been used to construct atomic bombs and to produce nuclear power through nuclear fission. In low doses, radioactive substances are generally not harmful—in fact they can be used as forces of good, as with X-rays that detect illnesses in early stages. But in higher doses radiation can kill you.

Doses of radiation are measured by units called rads or rems. The measurements indicate how much radiation is being absorbed in a single gram of human body tissue.

According to Atomicarchive.com, when you go to the doctor or dentist for an X-ray, you receive a dose of 1 rem of radiation. The bigger the dose, naturally, the more destruction is done to the body. If you were exposed to 100 rems of radiation, your "blood's lymphocyte cell count will be reduced," and you will become "more susceptible to infection." In other words, you will more easily be made into an actor capable of a non-SAG *28 Days Later* appearance.

At 200 rems you'd experience symptoms similar to a wretched flu—vomiting, nausea, diarrhea—a weird unknowable desire to watch late-night reruns of *Cheers* while pitying yourself and fantasizing that you are more like Sam than Norm, and also your hair might fall out in clumps. Thus, you will end up with the zombie look as characterized in chapter 3, rather than the one on the cover of this book (the cover zombie has probably been subjected to closer to 150 rems of radiation).

Anything over 400 rems can be fatal. Above 1,000 rems and the small blood vessels inside you are pretty much instantly toast. Heart failure is more or less a certainty at this point, and death overtakes you. By then your only chance for zombification is if some deranged lunatic found in a film like *Return of the Living Dead* or an evil Haitian witch doctor digs up your grave and resuscitates you to serve in his zombie army.

Having these numbers in mind, we can put the worst nuclear power catastrophe ever seen in its proper, malevolent context. After reactor number four at the Chernobyl Nuclear Power Plant exploded during a maintenance test early on the morning of April 26, 1986, emergency workers who heroically rushed to the site to help were exposed to about 20,000 rems per hour. The radioactive fallout from the accident has been estimated to be roughly 400 times greater than what the bombs dropped on Hiroshima generated. Amazingly, considering this destructive capacity, the city of 50,000 citizens saw a total of 30 people die, either from the initial fires or from the fatal levels of radiation. (Some 330,000 residents in the surrounding area were evacuated the morning after the reactor's cooling system malfunctioned—very few of them have returned.) But according to a 2006 story published in *National Geographic* magazine, the aftereffects of the disaster, the toxicity and disease wrought that morning in the northern center of Ukraine, will ultimately account for the deaths of 4,000 people.

Many of those deaths will be due to the cancerous mutation of genes brought on by the radioactivity of several isotopes, including cesium 137, strontium 90, and iodine 131. It is the iodine especially that many experts think has led to a massive uptick in cases of thyroid cancer among children living in proximity to the nuclear plant in Ukraine, Belarus, and Russia. Though the material has a relatively short half-life of eight days, it seizes aggressively on the thyroid gland and can lead to many severe medical problems.

The journal *Lancet* determined that children who lived through the Chernobyl explosion "may be more likely to develop hypothyroidism." That condition is associated with psychiatric manifestations identified also in schizophrenia. In fact, it can be difficult to distinguish between the two at times: while schizophrenia, for example, can cause general confusion and disorientation in patients, hypothyroidism can lead to stunted cognitive functioning. Hypothyroidism has also been linked to psychotic behavior and depression, which means patients suffering from this illness have been known to have a decreased cognitive rate, demolished attention span, less interest in human interaction, and insomnia. They are more likely to anger easily and in extreme cases patients might experience hallucinations of both sight and sound. Many of these traits, it is impossible not to notice, are shared by zombies: the short temper, yes, but also the lowered attention span, the insomnia, and the drifting toward misanthropy.

Taking the connection between hypothyroid disease and zombie-ism even further, we can take note of the thyroid gland's position in the system of hormone-controlling glands we all know as the endocrine system. This system includes the pineal gland, the thyroid, the thymus, and the pituitary. The pineal gland is a tiny sucker about the size of a kitten's toenail that sits in the brain not far from the cerebellum. It primarily produces melatonin, which, as any student who has ever crammed for a final can tell you, is a hormone that helps regulate our patterns of wake and sleep—our circadian rhythms, if you will. The thyroid is located in the neck and controls the production of hormones that deal with the body's metabolism and calcium levels. All the slacking thymus does, meanwhile, is develop T cells, which basically keep our bodies healthy.

That brings us to a fascinating little creature called the pituitary gland. Weighing in at a mere half a gram and about the size of a modest piece of bellybutton lint, the pituitary might not look like much on paper. Oh, but you'd be unwise to underestimate the brilliance of this hard-punching bantamweight of a gland. This baby secretes growth hormones, helps keep tabs on blood pressure, and dabbles in metabolic issues, too. That's when it isn't chipping in to keep our sex organs functioning properly and overseeing the efforts of the thyroid gland. So if you mess with the thyroid, you are, in essence, also messing with its anatomical bodyguard, the pituitary. And if you mess with the pituitary and cause it to go haywire, you could wind up with a body that is oversized, hulking, and abnormally strong. Just like a zombie.

One last thing about the pituitary and its zombific link: that's the gland that along with the brain's hypothalamus helps connect the body's nervous and endocrine systems. This means that in addition to being a close cousin to the thyroid, the pituitary gland is intimately familiar with the hypothalamus, which, as we learned in chapter 1, helps us know when we have had enough to eat. At least one Harvard psychiatrist we've mentioned in these pages, Dr. Steven Schlozman, believes that a zombie's insatiable taste for more, more, more flesh and brains suggests that we may be dealing with a distressed ventromedial hypothalamus (VH). And if the old VH is hurting, well, it seems possible that the pituitary gland might be acting up as well. Good-bye functioning nervous and endocrine systems, hello zombies!

It is not only in the ways that radiation affects us, changes us, mutates our genetic composition that resonates with the zombie experience. It is also the ways in which the radiation does not change us, cannot defeat us that we see a link with the undead. To explore this notion, we must return once again to Chernobyl. For it was there that citizens who'd been fed iodine pills and were bused out of town the morning after the unprecedented reactor meltdown—it would burn for ten days, producing toxic rain as far away as Japan—returned to live, in some cases mere weeks after the disaster. And while there have certainly been some health issues for the estimated 337 people who have moved back to the immediate area, for the most part their lives went on just as they did before April 26, 1986.

In 2006, *National Geographic* interviewed one such couple, Anna and Vasily Yevtushenko. That year Anna was seventy years old and Vasily was sixty-six. They lived in a tiny village called Opachichi along with seventeen other hearty souls prior to the nuclear explosion. Two years after having been evacuated from their home following the accident at the power plant, the Yevtushenkos decided to return to the town they loved and missed. So far, the incredibly severe risks involved with the Chernobyl horror seem to have had little or no effect on the Yevtushenkos. "If there was something, we would have already died," Vasily told *National Geographic*.

Now, either the couple is incredibly robust and also perhaps a tad lucky not to have been affected by the worst nuclear power debacle in history or something else is at work here. One seldom floated idea is that much like zombies who appear to be immune from such things as excessive radiation (being already dead and all), the Yevtushenkos simply carry on. They appear unmoved by the historic tumult around them and indeed go about their lives with a sort of mindless, zombielike consistency.

"I get up in the morning, I have chores to do," Anna says. "That's all." If you were expecting her to say she does that and, well, also craves madly for human flesh, you are not alone.

Should Anna and Vasily wander into the nearby town of Pripyat, they may yet develop additional undead characteristics. It is in Pripyat that a reporter's description of the post-Chernobyl scene could serve as the opening pages of any number of zombie screenplays:

> After passing through a checkpoint with a red-and-white-striped gate to deter looters and the merely curious, I meander through what was once a tidy town. Rows of white and pastel apartment blocks stand vacant, their windows dark and their lower stories overgrown. Near a kindergarten and a sports complex with a swimming pool, now empty and debris-strewn . . . stands a rusted Ferris wheel, its yellow cars groaning in the wind. It had been built just in time for May Day 1986.

It doesn't take too much of a stretch to believe that that passage comes instead from a new horror flick: *28 Weeks in Zombieland.*

While Anna and Vasily are among the few resettlers—or *samosely* as they are known—the animal kingdom near Chernobyl has seen a boom in postapocalyptic creatures. *National Geographic* catalogued the prospering wildlife not far from ground zero:

> More than a hundred wolves prowl the forest, endangered black storks and white-tailed eagles nest in the marshes, and several dozen Przewalski's horses, a rare breed that went extinct in the wild decades ago, are thriving after being released here in 1998. Pines are even reclaiming the Red Forest, though patches of

lingering radioactivity have left them stunted and deformed, with unnaturally short or long needles and clusters of buds where normally there would be just one. This radiation-warped forest is an anomaly. On the whole, ecologists marvel at how resilient nature has proved to be in the face of radiological adversity.

One such marveling expert was radioecologist Sergey Gaschak, who in 2006 told BBC News: "Animals don't seem to sense radiation and will occupy an area regardless of the radiation condition." Gaschak noted that while there was certainly still plutonium in the area (and with a half-life of nearly 24,500 years there would be for quite some time), the land surrounding the power plant was bereft of other things that might threaten an animal's existence: herbicides, pesticides, and automobiles. Wild boars—wolves be damned—were thriving!

Intrigued, Gaschak captured and tested creatures that had returned to the area. While he observed that there had been significant mutations in the animals' DNA, they were not terminally sickened or unable to reproduce. Or, as he put it to the BBC, "Nothing with two heads." (Well of course not—everyone knows zombie pets don't have two heads!)

So far in this chapter we've looked at the ways in which radioactive materials, just like in George Romero's first zombie movie, can transform us into DNA-twisted versions of ourselves. We've also seen how humans are capable of summoning enormous zombielike strength and courage in overcoming the harshest of obstacles—such as the toxic fallout at Chernobyl—to go about the endless cycles of our lives.

A third way that a conversation about radiation brushes up against an examination of the undead is the way in which the very radioactive isotopes themselves share qualities with vicious, hell-bent ghouls. The best example of this can be found in the history of uranium mining in America, in particular the relationship between uranium and the Navajo tribe of Native Americans. The bulk of that history takes place between World War II and 1971, though the effects of the mining are still reverberating across the country.

While uranium mining had been done in parts of Europe for decades prior to when it began in the United States, it wasn't until the 1940s that American industries began extracting the valuable and potent compound out of the earth in earnest. They did it primarily in the southwest in New Mexico, Utah, Colorado, and Arizona, and they did it with gusto: 1958 alone produced a haul of 7 million tons of ore; around this time there were roughly 750 active U.S. mines.

Many of those mines were worked by members of Navajo Nation, a southwest-based group of Native Americans who today live on some 16 million acres of land and represent the largest population of any Native American group in the world, with about 255,000 members. By one count, between 1945 and 1988 13 million tons of uranium ore was discovered on Navajo property.

Uranium has a well-known destructive capability, as it was a prominent part of the construction of atomic bombs. It is a wickedly powerful substance—the uranium 235 isotope is the only such material found in nature that can be enriched to the point that it is capable of a nuclear fission chain reaction.

By the time the stuff was being mined in the United States, a strong link between exposure to uranium and lung cancer had been established. One study conducted in 2000 found that there were ninety-four Navajo deaths due to lung cancer between 1969 and 1993, and that sixty-three of the victims had been uranium miners. Despite the early and overwhelming evidence of this deadly connection—a 1930 study done in Czechoslovakia made the claim, and a 1951 report in the United States detailed how radon preys on the lung's "sensitive cells for periods of time as long as their radioactive half-lives, delivering high doses of radiation"—it wasn't until 1990 that the Radiation Exposure Compensation Act passed, thus validating the science with legislation.

Is it any wonder then that Navajos consider uranium a monster? In fact they had two words for the beast: *Leetso*, or "yellow powder," and the Navajo word for "monster," *nayee*. According to the book *The Navajo People and Uranium Mining*, if you were to translate *nayee* literally into English, it would mean "that which gets in the way of a successful life."

As far as Navajo miners were concerned, the disease-spreading monster that most got in the way of a successful life was the *nayee*. And as far as Will Smith in *I Am Legend* was concerned, the disease-spreading monster that most interfered with a successful life was a zombie. Both monsters may appear innocuous at first—uranium is mined just as gold is, and it sparkles, too, while the virus-carrying zombies Smith battled were just one immunization away from being human—but they could and would kill you. This analogy is not intended to make light of the tragic circumstances surrounding

Navajo (and other) uranium miners of the twentieth century, but rather to show that evil can take many forms and monsters are lurking everywhere.

In 1968, intense fears about toxic junk that could ostensibly find its way into our atmosphere—by way of a busted satellite's radiation—gripped us with the strength of ten zombies. Today we live in a world that makes the late sixties look like Mister Rogers' Neighborhood. The Earth is heating, the ozone is melting, and the airwaves are burning with vitriol. If we aren't done in by the sickening spritz of pesticides, cell phone–induced ear cancer will get us. Our mobile devices are techno time bombs ticking toward a pocketful of wireless pain. Smartphones may be smarter than us and, if so, are already calculating how to take over the world after a cryogenically frozen Steve Jobs is done being president. Science fiction is now filed under current events.

All of this suggests that the contemporary threat of a zombifying radioactive episode appears not just possible, but inevitable. Here we've looked back at history to see how such disasters have played out. We would have looked at the future, too, but we don't have a toxic-spill-provoked, DNA-mutated, all-seeing cyclops eye (yet!). And so the future remains unknown, hiding in the shadows of time unspent like something that might get in the way of a successful life. Should the future bring an Earth-destroying radiation disaster—after which only a few unlucky survivors must flee prowling packs of zombies, running, running, running while keeping an eye out for a motorboat—that would be terrible. But at least now it won't catch any of us totally by surprise.

YOU GOTTA SHOOT 'EM IN THE HEAD!

...

On ballistics, physics, recoil, zombies, and you.

■ ■ ■ ■ ■

BY NOW we all have a pretty good idea what we're talking about when we talk about a scientific investigation of zombies. We know about their neurobiological makeup (mostly just angry); we know about the chances any of us has of getting all zombied up due to a viruslike outbreak (alarmingly high unless we start fighting back *rightthisminute*); we even know how zombies get so big and strong (lots of exercise and plenty of delicious flesh 'n brains!).

One remaining question lingers, however, and it's a real doozy. That's right, I'm talking about the best way to blow their freakin' heads off. Ironically, pretty much the only way to defeat the suckers is by dismantling *their* brains—as this is arguably the only functioning organ inside the lumbering undead—so I think it would be helpful to take a look at the ideal tools for the job. To understand the available choices, let's dig into a discussion of ballistics and the physics of phucking things up.

Countless films—as well as science, as we shall see—make it clear that some form of firearm is the way to go in terms of efficiency, so we won't even dwell on weapons like nunchucks, tanks, grenades, machetes, and frying pans. Let's go straight for the serious ammo. I can't emphasize enough just how crucial this information will be to your future well-being.

In fact, don't take it just from me; *The Zombie Survival Guide* author Max Brooks has some very strong opinions on the matter as well: "Of all the weapons discussed in this book, nothing is more important than your primary firearm. Keep it cleaned, keep it oiled, keep it loaded, keep it close. With a cool, steady hand, and plenty of ammunition, one human is more than a match for an army of zombies."

I like Brooks's confidence there—and he certainly makes survival sound plausible—but unfortunately it's not as simple as all that. A few pages later, in fact, Brooks himself notes that even the best laid plans (and fastest firing guns) can go south in the heat of the moment. He writes:

> Studies have shown that, given the trauma of battle, the closer a human is to a zombie, the wilder his shooting will be. When practicing with your firearm(s), establish a maximum range for repeated accuracy. Practice against moving targets in ideal (stress-free) conditions. Once that range is fixed, divide it by half. This will be your effective kill zone during an actual attack.

X

That's good advice. But first we must determine which weapon is best—not just for you, but for mowing down zombies under duress.

Sometimes it helps to have a goal in mind—something to shoot for, as it were. May I suggest listening to Robert A. Rinker while thinking about your goal? Rinker, after all, wrote one of the best books on ballistics out there, *Understanding Firearm Ballistics: Basic to Advanced Ballistics Simplified, Illustrated, and Explained*, and offers a tried-and-true method for stretching your firing skills as far as they can go. Rinker says: "The ultimate in precision shooting is the 1,000-yard range, and the accuracy obtained is nothing short of astounding. The best shooters can obtain groups of around 4.25″ to 4.75″ with 10 shots at 1,000-yards." In other words, he suggests you place a small paperback book (not this one!) against a tree and drive .55 miles away from the tree. Once you are able to shred the pages of that book, you are more than ready to face off against the undead. Your shots will have landed in that range he spoke of, all clustered within the same half an inch of one another. Keep in mind, however, that most of the time you are firing on zombies it will be at distances far shorter that 1,000 yards. So while this is an admirable goal, it is not yet the most crucial skill to pursue.

For the amateur zombie fighter, Rinker is a good place to start, blitzing as he does through so many essential ideas on ballistics. To start with, he says it's important to keep in mind Sir Isaac Newton's three laws of physics:

1. A body in motion is just going to continue in motion in the exact same way, with the exact same speed unless

something comes to disrupt it (like a zombie getting in the way of your bullet).

2. Force = mass × acceleration. This is important when calculating what kind of damage you and your gun are about to do to a zombie.
3. For every action there is an equal and opposite reaction. The hammer of your gun hits powder; your ammo is launched in the direction of a zombie.

These are the primary principles driving most conversations about ballistics. They inform the necessary weight of your weapon, the speed of the bullets, and the amount of recoil (what most people think of as kick) you can expect to feel in your shoulders and arms. In many cases, expect to feel it even the morning after you shoot. Weight and recoil are important considerations, especially when you know that you may be lugging your gun for a long, long time and will likely be firing it often.

Brooks has his eye on one weapon that might make for the perfect zombie-fighting accoutrement—the semiautomatic rifle. "Since its debut," Brooks says, "this weapon has shown itself to be a superior zombie killer," and he even cites one zombie battle where "a trapped woman dispatched fifteen attacking zombies in twelve seconds!" Without a doubt, those are impressive stats. But this type of rifle might not be for everyone. There are many factors involved when it comes to ballistic efficiency, which Rinker says "is figured by the ratio of pressure to velocity." Among these factors are the bullets' weight, shape, and jacket core, a gun's loading density, and the temperature of the cartridge and gun when fired. On this last point, Rinker observes that heat, brought on by exploding gases from the powder, "can damage a barrel and change accuracy." For this reason, he says, machine gun barrels have an extremely short life span because they generate a lot of barrel-warping heat.

It is also for this reason that Rinker recommends taking your time between zombie shots, though that may be hard to stick to once zombies are shuffling on your trail. "For accuracy," however, "slower shooting requires the shots be spaced to hold a uniform barrel temperature, perhaps a little warm but not too hot." The weaponry wizard gives us a chilling thought: "Remember that the special, once in a lifetime shot, will almost certainly be from a cold barrel." This advice seems most practical if you are fending off a single ghoul; should you find yourself under assault by a swarm of flesh seekers, a liberal application of Brooks's semiautomatic rifle might be a better bet.

Many zombie hunters are partial to shotguns or even double-barrel shotguns (think Ash in the second and third Evil Dead movies, among many others). This makes sense as shotguns look really cool and also have what is known as stopping power, the ability to prevent the undead from continuing the chase. What you should know before you rush out and buy up all the shotguns you can find is that they produce what can be a shoulder-snapping recoil. Given the heft of the gun and the speed it can generate, Rinker says that "firing both barrels simultaneously from a double barrel shotgun is not one of life's little pleasures. Yet, if the gun is held firmly and properly, the shooter should not receive any permanent injury."

The fact that permanent injury is an option should give double-barrel fans pause. Here are some hard numbers to help your decision, however: "The average double shotgun will weigh about 7.25 lbs and with a 3.75–1.25 load, will have 34.5 ft. lb. recoil. If both barrels fired together, the energy would be 138-ft. lbs." ("Ft. lbs." is the measurement used for talking about the energy of a shot as it leaves the gun. One foot-pound means that 1 pound of force is moving across 1 foot of distance.) That's four times as much force than that from a single barrel because you are dealing with twice as much powder and ammo squared. Here is a basic formula for assessing recoil in general:

$$MG \times VG = MC \times C + MB \times BV$$

And the key for that is:

MG = Mass of gun
VG = velocity of gun
MC = mass of powder charge
C = velocity of powder charge/ the expanding gases, exiting the muzzle
MB = mass of bullet
BV = velocity of bullet

You can easily determine all of these factors when deciding on your weapon, and then with this equation you can determine how much recoil is right for you.

If you are on the go (and I'm guessing you are), a simpler way of calculating recoil is to use only the weight of the gun and the weight and velocity of the bullet. Rinker gives the example of a 9-pound gun—admittedly kind of a heavy mother. Since bullets are weighed in grains and there are 7,000 grains to a pound, we see that the gun weighs 63,000 grains. A typical bullet for a gun this size is 150 grains, meaning the gun is 420 times greater in weight than the bullet. Rinker states that the recoil of this combination will be "1/420th of the bullet's velocity."

Recoil is so important to the discussion of ballistics because it can greatly influence accuracy, which, for our purposes, is really the most meaningful gauge of a weapon's worth. With that in mind, here are a few other helpful tips per Rinker:

- An ordinary rifle of .30 caliber with normal loads will recoil back about .060 inch before the bullet leaves the muzzle.
- The lighter guns cause less fatigue during a long hunt where walking is involved. They also kick more. What one gives, the other takes away. One consolation is that most hunters are under enough excitement at the moment of kick that it goes almost unnoticed.
- A great rifleman . . . said that most shooters could handle up to about 15-ft. lbs of recoil without problems. Guns in the class of the .458 Winchester, which has a 60-ft. lb. recoil, are not for everyone. [Think of what Shaun wielded in *Shaun of the Dead*.]

And so the question still stands: which gun, given the task at hand (killing zombies), is right for you? One answer is: whatever gun you can comfortably fire from long distances and with great accuracy without suffering permanent injury from recoil. The toughest among you may actually be able to wield the aforementioned Winchester, while others should probably consider Brooks's semi or perhaps a double-barrel shotgun if the outbreak is not yet too severe and does not require constant firing. Another answer involves an examination of ammo, which we will get to in a moment.

Before we leave the gun show, however, a couple of notes are needed on pistols. These concealable little numbers certainly have a place in the arsenal of any serious zombie assassin. Yes, they have their flaws, especially when it comes to stopping power. Brooks claims, "Studies have shown that of all wasted ballistic wounds—e.g., those that struck a zombie in a non-lethal way—73 percent came from some type of handgun." Still, he allows, "The fact that handguns are small, light, and easy to carry make them attractive as a secondary weapon for any scenario."

Rinker, naturally, has a detailed breakdown on this very subject. One important thing to keep in mind is his statement on accuracy: "All else being equal, a handgun will shoot lower on the target with light loads and higher with heavy loads no matter the velocity . . . the reason is with barrel jump and gun movement [associated with] interior ballistics." In other words, different bullets have different weights, which can affect what is happening inside the barrel and, subsequently, what happens to the gun as you fire.

Rinker's thinking on handguns comes down to a comparison between revolvers and semiautomatic pistols. He gives the pros and cons of each in this way:

- Pro: "The semi-automatic will usually carry a large amount of cartridges and can be reloaded faster with a clip . . . the semi-automatic will fire faster." Advantage semi. But not so fast!

- Con: "The safety on a semi-auto can work both for and against its owner. If it is forgotten in an emergency, then the weapon may not fire when needed. If the gun is taken away and turned against him or her, the opponent can forget the safety or not know how to release it. This may save the owner's life."

This calls to mind the scene in *Day of the Dead* when the smarty-pants zombie Bub finds himself in possession of a handgun and uses it most effectively on the humans holding him hostage. Bub, it would seem, can find his way around a safety.

The four main types of bullets available to the general gun-shooting public are

- Nonexpanding bullets that are built to retain their shape as they pass through their targets. Their nose is covered with a second metal that's lighter than lead, and the body of the bullets is lead and uncovered, which is good for hunting as they don't shred the animal.
- Expanding bullets that are meant to expand on impact and create a wound with a larger diameter while not penetrating too deeply. They're built with a softer nose that shatters on impact, causing a severe wound.
- Fragmenting bullets that are designed to break apart. These bullets have a hollow or extremely soft nose and practically turn to tiny metal dust inside the target.
- Partially fragmenting bullets, which are hybrid bullets with a nose that is soft and breakable and a hard body able to penetrate deep into the target.

"The importance of the bullet to the performance of a firearm cannot be overemphasized," Rinker writes. For our purposes—shooting zombies in the head—we are less concerned with what happens after the bullet hits than before. That is, we need to know what the weight, the shape, and the material of a bullet do for speed and accuracy. There are many factors.

A gun's caliber, for example, refers to the diameter of the interior of the barrel, or what is called the bore. But it can also be the measure of the diameter of the bullet. So a .22 caliber rifle has a measurement of .22 inches in diameter in the bore inside the gun's barrel. A cartridge is a full round of ammo ready to go with everything you need in that one container: the case surrounding the cartridge (made of anything from brass, copper, and steel to plastic or even paper), the powder, and a chemical compound propellant, which helps, you know, propel stuff and the bullet. The shell shot from a shotgun could be considered the same thing as a cartridge but is usually instead called a shotgun shell. In short: shooting zombies is not easy!

In general, smaller bullets are lighter and faster but less likely to stop a zombie. Larger bullets are heavier and slower and can do more damage. You must also choose from a variety of materials—lead or tin usually for the bullet's body and copper or steel for the thin tip or jacket surrounding the projectile's core.

All in all, Rinker makes a strong case for the boattail—or tapered—design for bullets. These are shaped like miniversions of rocket ships. The benefits of this design—speed, mostly, measured as feet per second (feet/second)—increase as the distance increases. This is because "there is a partial vacuum behind a bullet. This partial vacuum creates base drag that helps slow the velocity. The vacuum is less behind a boattail bullet than a blunt end." Rinker elaborates on the upside of this type of bullet. "As the range increases, the benefit increases. At 1,000 yards the velocity difference can be improved by as much as 20% with up to a 3 or 4 foot advantage in wind deflection in a modest 10 mph cross wind. The boattail design has a much longer extreme range than a flat-based bullet if the velocity and weight are the same." And there's that magical 1,000 yards again. Ten football fields lined up in a row. It might seem an impossibly long distance at which you will want to pop a cap in a zombie forehead, but wouldn't it be nice if you had that option and didn't need to wait until it was within handgun range?

The spitzer design for bullets also produces impressive results. This design emphasizes a longer, leaner look, and its aerodynamic shape helps it maintain a high velocity over long distances. If you were to fire, at the same velocity, a spitzer bullet equal in weight to a round-nosed bullet at 300 yards, according to Rinker, "the spitzer has an advantage of 300 feet/second in velocity and 475 foot-pounds of energy, an impressive difference."

So far we have remained in the realm of the theoretical—what if a zombie attacked; what would the best choice of weaponry be in that case?—but it would be great to get some insight from someone who has had to deal with these issues in the real world. Luckily, Jonathan Maberry found just such a fellow to comment on this subject for his book *Zombie CSU*. The gentleman Maberry sought wisdom from was a member of the South Central Special Emergency

Response Team in Bucks County, Pennsylvania, Sergeant Ted Krimmel. Krimmel says that he and his men "often train officers to shoot center mass," as the middle of the body is "the easiest thing to hit in a high stress situation." That's good thinking, but still we need to aim for the head, don't we?

"Assuming the premise that the zombie will only be killed by a head shot," Krimmel says, "a shot to the chest could potentially cause enough blood loss to at least slow a creature down." Okay, that could work. And the best weapon for the slowing job? "A 12-gauge shotgun loaded with 00 buckshot (standard police load) fires 9 .33 caliber pellets at once. The tissue damage and resulting blood loss at close range is devastating. It should have enough energy to enter the chest and exit the rear of the suspect."

Other options Krimmel recommends for doing similar damage to the encroaching ghoul are an M-4 or assault rifle loaded with 5.56 mm tactical bonded bullets, where the core and jacket are soldered or otherwise tightly connected so that they will not separate at any time, including at impact.

If you find yourself deciding between an assault rifle, like the much-loved AK-47, or a 12-gauge shotgun, the Remington 870 say, you can rest easy knowing there really isn't a bad choice to be made. Some other information that may help your decision: the rifle weighs around 6.8 pounds while the shotgun is closer to 8. That might seem like a negligible difference, but that pound plus adds up fast when you are trekking through acre after acre of wooded land, fear pulsing in the hard vein of your head.

Perhaps more important, keep in mind that the shotgun is accurate up to around 50 yards, while the assault weapon laughs at 50 yards and can be effective up to 300 yards, sometimes even more. If you expect to be in the trenches with the undead, a 12-gauge should be fine; if you would like to keep your distance, consider the rifle.

You may not have to decide between these two zombie stoppers at all. If you listen to Max Brooks, your stockpile of essential firearms will include a rifle, a shotgun, *and* a handgun. (That's in addition to a crossbow, a sword, two knives, a hatchet, and more.) This seems about right for home use, but what about when you are out there on the road, running, leaping, hiding, and getting it on with a punked-out Emma Stone? What then? You will want not only significantly fewer gats to lug around than Brooks proposes, but also probably the lightest zombie stopper on the market.

When I add up all of the analyses from our impressive collection of experts on the subject, the answer I arrive at may not surprise you: I'd go with a light but super sturdy semiautomatic assault rifle, as Brooks suggested. By all accounts the AK-47 is still at the top of the class. Use 180-grain spitzer-shaped bullets for maximum velocity and accuracy over long distances. You will probably also want to have a semiautomatic handgun shoved into your waistband just in case—and when it's go time, make sure the safety is off. The 9 mm Glock is a popular choice. Yes, handguns often require more shooting practice than revolvers and are prone to jam, but they also fire much faster than revolvers, which is something you will find very helpful when the undead are closing in on you.

I must say it feels a little funny to try to offer utilitarian advice here at the end. For the most part this text is meant to help us all simply better understand the troubling beasts that would devour us. Personally, I don't see a lot of value in studying how to defeat zombies through sheer stockpiling of weaponry. That seems a fool's errand. No, I would say a superior approach would be to study the creatures now and hope to befriend them later. Defeating them is not really an option.

Anyway, have you ever noticed how everyone seems so miserable in those zombie movies? Is that the life you want for yourself? Paranoid, on edge, wits shot, terrified, and stuck in the same shapeless, tattered, and filthy clothes day after day. Is that the best the future holds?

Sure, go out and arm yourself for the undead invasion (once the invasion commences all of the salespeople at gun shops will have either fled or been turned into zombies, so you can have your guns for free!). But don't fight too hard. It's just not worth it. You will go down eventually. Instead, I say choose life (or, rather, being undead) after death. Think about it—no more taxes, illness, lines at the grocery store, or traffic on the expressway. Now everywhere you look, it's an all-you-can-eat buffet! If that's not the American Dream, I don't know what is.

Mad Limbs #2

■ ■ ■

Best if read by an athletic heartthrob wearing at least one item of ripped clothing. Someone with a bleeding wound—bad, but not bad enough to stop him from being abnormally good looking. A natural leader with just enough toughness and hair that cannot possibly look bad and eyes like the sea and a dark secret that isn't really that dark—that he can confess in a moment when death seems certain, just before he has an idea that will save everyone. Must be able to drive both stick and automatic.

Youngish Leader Guy

Okay, this is it. I don't know how long that lock's gonna hold.

Maybe ____________ (NUMBER) minutes, maybe ____________ (NUMBER).

Now listen to me. We've all lost a lot; God knows we have.

But it's gone. The past doesn't matter anymore. It doesn't

matter that I ________________ (VERB, PAST TENSE) that ________________ (SPECIES OF BIRD)

in ________________ (LOCATION). ________________ (MALE NAME), nobody here

cares that you sold ________________ (ILLICIT SUBSTANCE) in an alley behind

the ________________ (RESTAURANT). ________________ (FEMALE NAME), you

____________________, if you hadn't ____________________
ADJECTIVE NOUN VERB, PAST TENSE

when that zombie ________________________ was trying to
EPITHET

______________________ us, we'd all be dead. All of us.
VERB

And so what about the past? Everybody who cares about the past is a goddamn zombie now.

All we need to do is get to that ____________________. If we
VEHICLE

can do that, I swear on _____________'s __________________
DEITY BODY PART

we'll all get out of here and make it through this.

And when people look back and write about us, about what we did tonight, they'll say maybe those guys were

_________________, and maybe they had _________________,
ADJECTIVE DISEASE

but ________________________! They fought those zombies
EXPLETIVE

when it counted. Those guys grabbed redemption by the

______________________. That's what they'll say.
BODY PART

This is our time! ____________________, on three you throw
MALE NAME

the __________________ and we all run for it. One . . . two . . .
NOUN

three, ____________________!
EXCLAMATION

NOTES

1. Know Thy Enemy

16 *five authors, three of whom* Mark Strauss, "A Harvard Psychiatrist Explains Zombie Neurobiology," io9.com, http://io9.com/5286145/a-harvard-psychiatrist-explains-zombie-neurobiology.

18 *the major points of Dr. Schlozman's paper* Steven Schlozman, "The Neurobiology, the Psychology and the Cultural Overtones of the Zombie Film Genre," Phlog podcast audio program, *Boston Phoenix*, April 16, 2009, http://thephoenix.com/BLOGS/phlog/archive/2009/04/16/podcast-harvard-psychologist-explains-the-science-of-zombies.aspx.

18 *"clinically speaking, impulsivity is"* Ibid.

19 *"you can't really be mad at zombies"* Ibid.

20 *"Frontal lobes and amygdala are always talking"* Ibid.

20 *"If you've ever stayed out late drinking"* Ibid.

20 *"Ataxia often occurs"* "NINDS Ataxias and Cerebellar or Spinocerebellar Degeneration Information Page," National Institute of Neurological Disorders and Stroke, www.ninds.nih.gov/disorders/ataxia/ataxia.htm.

21 *"we are wired to connect"* Schlozman, "The Neurobiology, the Psychology and the Cultural Overtones of the Zombie Film Genre."

22 *"What if the enemy can't be shocked"* Max Brooks, *World War Z: An Oral History of the Zombie War* (New York: Random House, 2006), 104.

23 *that day Gage was overseeing a project* Hanna Damasio, Thomas Grabowski, et al., "The Return of Phineas Gage: Clues about the Brain from the Skull of a Famous Patient," *Science* 264 (1994): 1102–1105.

24 *"But he survived a different man"* Ibid., 1102.

24 *"in some respects, Gage was fully recovered"* Ibid.
24 *"On the other hand, he had become"* Ibid.
25 *"the equilibrium or balance"* Ibid.
25 *"Gage was no longer Gage"* Ibid.
25 *"It may be that pathological aggression results"* Pamela Blake and Jordan Grafman, "The Neurobiology of Aggression," *Lancet* (2004): 2–3.
26 *"With normal growth and development"* Ibid., 2.
27 *when the photos were shown for longer* "Brain Activity Reflects Complexity of Responses to Other-Race Faces," Science Daily, www.sciencedaily.com/releases/2004/12/041208231237.htm.
27 *"the essence of zombies is amygdala"* Schlozman, "The Neurobiology, the Psychology and the Cultural Overtones of the Zombie Film Genre."
27 *the region of the brain* "Less Empathy toward Outsiders: Brain Differences Reinforce Preferences for Those in Same Social Group," Science Daily, www.sciencedaily.com/releases/2009/06/090630173815.htm.
28 *"If there were machines bearing"* Rene Descartes, *Discourse on Method* (Upper Saddle River, NJ: Prentice Hall, 1956), 33.
29 *"The second test"* Ibid.
29 *"Again, by means of these two tests"* Ibid.

2. Serve with a Chilled Pinot Gross

35 *"It is a truth universally acknowledged"* Seth Grahame-Smith, *Pride and Prejudice and Zombies: The Classic Regency Romance—Now with Ultraviolent Zombie Mayhem!* (Philadelphia: Quirk Books, 2009), 7.
36 *discovered a genetic code suggesting* John Collinge, Simon Mead, et al., "Balancing Selection at the Prion Protein Gene Consistent with Prehistoric Kurulike Epidemics," *Science* 300 (2003): 640–643.
38 *"free the spirit of the dead"* Mike Alpers, "The Epidemiology of Kuru: Monitoring the Epidemic from Its Peak to Its End," *Philosophical Transactions of the Royal Society B* 363 (2008): 3707–3713.
38 *"a fatal neurodegenerative disease"* Ibid., 3707.
39 *"strong evidence for widespread cannibalistic practices"* Collinge, Mead, et al., "Balancing Selection at the Prion Protein Gene Consistent with Prehistoric Kurulike Epidemics," 640–643.
40 *"there is no animal prion disease"* Nicholas Wade, "Gene Study Finds Cannibal Pattern," New York Times online, www.nytimes.com/2003/04/11/us/gene-study-finds-cannibal-pattern.html?pagewanted=1.

40 *cannibalism is not nearly as popular* Ibid.

40 *"starts with protracted insomnia"* "Prion Diseases," BBC News, http://news.bbc.co.uk/2/hi/health/medical_notes/355601.stm.

42 *"We used seal oil for cooking"* Patricia Gadsby, "The Inuit Paradox," Discover online, http://discovermagazine.com/ 2004/oct/inuit-paradox.

42 *"Presumably, from a nutritional standpoint"* E-mail correspondence with the author, December 15, 2009.

43 *When that same nutritionist crunched the numbers* Correspondence with the author, December 15, 2009.

43 *"surprising to learn how well"* Gadsby, "The Inuit Paradox."

43 *"no essential foods"* Ibid.

44 *"A pound of frozen walrus"* Edmund Searles, "Food and the Making of Modern Inuit Identities," *Food and Foodways: History and Culture of Human Nourishment* 10 (2002): 55–78.

44 *"It was not unusual for him to consume"* Ibid., 65.

44 *"Aksujuliak was known in Iqaluit"* Ibid.

45 *"[The] Inuit encourage their children"* Ibid., 72.

46 *"If you have some fresh meat"* Gadsby, "The Inuit Paradox."

47 *A 3.55-ounce chunk of raw caribou liver* Ibid.

47 *the Inuit diet as the original Atkins* Ibid.

48 *"These man-made fats are dangerous"* Ibid.

49 *"a zombie's digestive tract"* Max Brooks, *The Zombie Survival Guide: Complete Protection from the Living Dead* (New York: Random House, 2003), 11.

49 *"This partially chewed, rotting matter"* Ibid., 12.

50 *"One captured and dissected specimen"* Ibid.

3. Earth Worms Are Easy

60 *"the destruction of the soft tissues"* Arpad A. Vass, "Beyond the Grave—Understanding Human Decomposition," *Microbiology Today* 28 (2001): 190–192.

61 *in one experiment* "Decomposition: Forensic Evidence," Australian Museum, http://australianmuseum.net.au/Decomposition-Forensic-Evidence.

61 *"By seven days after death"* Kenneth V. Iserson, *Death to Dust: What Happens to Dead Bodies?* (Tucson, AZ: Galen Press, 2001), 51.

62 *"The skeleton also has a decomposition rate"* Vass, "Beyond the Grave," 190–192.

63 *"[zombies] are the recent dead"* Jonathan Maberry, *Zombie CSU: The Forensics of the Living Dead* (New York: Kensington, 2008), 13.

65 *"Are zombies truly dead"* Ibid., 96.

65 *"essential oils, aloe, salt, myrrh"* Iserson, *Death to Dust*, 276.

65 *unintentional embalming* Vass, "Beyond the Grave," 190–192.

66 *"had sterilized the body"* Ibid., 192.

67 *"A four-sided plate covered the abdominal"* C. Baldock et al., "3-D Reconstruction of an Ancient Egyptian Mummy Using X-ray Computer Tomography," *Journal of the Royal Society of Medicine* 87 (1994): 806–808.

67 *"The soft tissue was more radio-opaque"* Ibid., 807.

68 *During a series of tests in 2005* Rajiv Gupta, "High Resolution Imaging of an Ancient Egyptian Mummified Head: New Insights into the Mummification Process," *American Journal of Neuroradiology* 29 (2008): 705–713.

70 *"The physicality of a human corpse"* Iserson, *Death to Dust*, 52.

4. Sex and the Single Zombie

79 *"sterilizing trematode parasite"* Maurine Neiman et al., "Accelerated Mutation Accumulation in Asexual Lineages of a Freshwater Snail," *Molecular Biology and Evolution* 27 (2009): 954–963.

79 *"It turns out"* Maurine Neiman, e-mail message to author, February 1, 2010.

81 *"This is the first study"* "Value of Sexual Reproduction versus Asexual Reproduction," Science Daily, www.sciencedaily.com/releases/2010/01/100121161238.htm.

81 *"Having regular and enthusiastic sex"* Alan Farnham, "Is Sex Necessary?" Forbes online, www.forbes.com/2003/10/08/cz_af_1008health_print.html.

82 *There is a large postcoital increase* Ibid.

83 *A study done by a Pennsylvania university* Farnham, "Is Sex Necessary?"

5. Unsafe at Any Speed

92 *3.5 grams of it could render* William Booth, "Voodoo Science," *Science* 240 (1988): 274–277.

92 *"appeared comatose and showed no response"* Ibid., 275.

93 *"causal agent in the initial zombification"* Ibid., 276.

94 *"I've never maintained"* Ibid., 276.

94 *"psychiatrist and pioneer"* Wade Davis, *The Serpent and the Rainbow: A Harvard Scientist's Astonishing Journey into the Secret*

Societies of Haitian Voodoo, Zombis, and Magic (New York: Simon & Schuster, 1985), 22.

94 *He had a high fever* Ibid., 28.

95 *As the Web magazine Slate noted* Josh Levin, "Dead Run," Slate, www.slate.com/id/2097751.

97 *"Local beliefs about body, mind and spirit"* Glennys Howarth and Oliver Leaman, eds., *Encyclopedia of Death and Dying* (New York: Routledge, 2001), 679.

6. What They Don't Teach You in Health Class

106 *They published their results* Ioan Hudea et al. "When Zombies Attack!: Mathematical Modelling of an Outbreak of Zombie Infection," in *Infectious Disease Modelling Research Progress*, ed. Jean Michel Tchuenche and Christinah Chiyaka (Hauppauge, NY: Nova Science Publishers, 2010), 133–150.

108 *"when a susceptible individual"* Ibid., 134.

109 *"removing the head"* Ibid., 135.

109 *"This model is slightly more complicated"* Ibid.

109 *"Susceptibles first move to an infected class"* Ibid., 138.

110 *"humans are not eradicated"* Ibid., 144.

111 *"Only sufficiently frequent attacks"* Ibid., 146.

112 *"Even assuming the entire world is susceptible"* Matt Mogk, "How Zombies Can Destroy the World," Zombie Research Society, http://zombieresearch.net/2009/11/12/how-zombies-can-destroy-the-world/.

112 *"Not going to happen"* Jonathan Maberry, *Zombie CSU: The Forensics of the Living Dead* (New York: Kensington, 2008), 34.

113 *"If the plague does not spread"* Ibid., 137.

113 *"Even the common cold"* Ibid.

113 *"we could see a frightening pattern"* Ibid., 111.

116 *"the pandemic's most striking feature"* "Researchers Reconstruct 1918 Pandemic Influenza Virus; Effort Designed to Advance Preparedness," Centers for Disease Control, www.cdc.gov/media/pressrel/r051005.htm.

117 *all other flu pandemics are "descendants"* David M. Morens and Jeffery K. Taubenberger, "1918 Influenza: The Mother of All Pandemics," *Emerging Infectious Diseases* 12 (2006): 15–22.

117 *"composed of key genes"* Ibid., 15.

117 *"On the contrary"* Ibid., 18.

117 *"For example, the 1918 nucleoprotein"* Ibid.

117 *"All of these findings"* Ibid.

118 *"lower environmental and human"* Ibid., 17.

120 *"One of the most difficult"* Niall P. A. S. Johnson and Juergen Mueller, "Updating the Accounts: Global Mortality of the 1918–1920 'Spanish' Influenza Pandemic," *Bulletin of the History of Medicine* 76 (2002): 105–115.

7. Do Zombies Dream of Undead Sheep?

124 *According to the Centers* "Insufficient Rest or Sleep in Adults, United States, 2008," CDC Features, www.cdc.gov/Features/dsSleep/.

126 *"Partial sleep deprivation appeared"* Allen I. Huffcutt and June J. Pilcher, "Effects of Sleep Deprivation on Performance: A Meta-Analysis," *Sleep* 19 (1996): 318–326.

129 *"there can no longer be any doubt"* Allan J. Hobson and Edward F. Pace-Schott, "The Cognitive Neuroscience of Sleep: Neuronal Systems, Consciousness and Learning," *Nature Reviews Neuroscience* 3 (2002): 679–693.

129 *"So waking suppresses hallucinosis"* Ibid., 686.

129 *"Different regions are . . . hyperactivated"* Ibid.

131 *"Brain activity* in utero*"* Ibid., 688.

132 *"It's really an admixture of wakefulness"* From an interview with author.

135 *"I was exhausted, nervous"* Max Brooks, *World War Z: An Oral History of the Zombie War* (New York: Random House, 2006), 212

136 *"To sleep—perchance to dream"* William Shakespeare, *Hamlet* (New York: Simon & Schuster, 2004), 3.1. References are to act and scene.

8. Bee Afraid, Bee Very Afraid

140 *"He discovered that while parasites"* Carl Zimmer, *Parasite Rex: Inside the Bizarre World of Nature's Most Dangerous Creatures* (New York: Simon & Schuster, 2001), 27.

140 *"Parasites live in a warped version"* Ibid., 46.

140 *"Drinking blood is not easy"* Ibid., 90.

142 *"It's lost its will"* Carl Zimmer, interview by Jad Abumrad and Robert Krulwich, *Radiolab*, WNYC, September 7, 2009.

143 *"I think the most likely explanation"* Nora Schultz, "Zombie Cockroaches Revived by Brain Shot," *New Scientist*, www.newscientist.com/article/dn12983-zombie-cockroaches-revived-by-brain-shot.html.

144 *"Our brain is of course much more complex"* Ibid.

145 *"the most vicious and unneighborly behavior"* Zimmer, *Parasite Rex*, 48.

145 "T. gondii *is of special interest*" E. Fuller Torrey and Robert H. Yolken, "*Toxoplasma gondii* and Schizophrenia," *Emerging Infectious Disease* 9 (2003): 1375–1380.

146 *"Once* Toxoplasma *has invaded a cell"* Zimmer, *Parasite Rex*, 67.

147 *"If the parasite were to multiply"* Ibid., 68.

148 *"after an acute infection"* John C. Boothroyd et al., "Behavioral Changes Induced by *Toxoplasma* Infection of Rodents Are Highly Specific to Aversion of Cat Odors," *National Academy of Sciences* 104 (2007): 6442–6447 (emphasis added).

148 *"neuroanatomical circuit comprising"* Ibid., 6442.

150 *"The core of parasitism is the ability"* Ibid., 6446.

150 *"Some cases of acute toxoplasmosis"* Torrey and Yolken, "*Toxoplasma gondii* and Schizophrenia."

151 *"Neuropathologically, studies of* T. gondii" Ibid., 1378.

153 *The Leeds study found a possible link* "Research Supports Toxoplasmosis Link to Schizophrenia," University of Leeds, www.leeds.ac.uk/news/article/590/research_supports_toxoplasmosis_link_to_schizophrenia.

9. Fear and Loathing in Zombietown

160 *"Those who say they are affected"* Andrea Thompson, "'Meteorite' Crash Breeds Mass Hysteria," FoxNews.com, www.foxnews.com/story/0,2933,298177,00.html.

160 *"The Peruvian event seems to be a rare case"* Ibid.

160 *"In recent years, there have been numerous cases"* Ibid.

161 *"We don't know what is going on"* "Scores Ill in Peru 'Meteor Crash,'" BBC News, http://news.bbc.co.uk/2/hi/7001897.stm.

161 *"the phantom anesthetist of Matoon"* Leslie P. Boss, "Epidemic Hysteria: A Review of the Published Literature," *Epidemiologic Reviews* 19 (1997): 233–242. The phantom anesthetist of Matoon was one of a few nicknames bestowed upon the mysterious person or group that may have perpetrated a series of gas attacks in the 1930s in Botetourt County, Virginia. Authorities are still not certain if there actually were attacks or if it was a case of mass hysteria.

161 *"Too often, it is the media-created event"* Ibid., 237.

163 *"The stressful nature"* Ibid., 236.

164 *"Over 2,300 anthrax false alarms"* Robert E. Bartholomew and Simon Wessely, "Protean Nature of Mass Sociogenic Illness," *British Journal of Psychiatry* 180 (2002): 300–306.

165 *"later it spread to two"* L. P. Kok, "Epidemic Hysteria (A Psychiatric Investigation)," *Singapore Medical Journal* 16 (1975): 35–38.

165 *"saw a shadow in front of her"* Ibid., 36.

166 *"overly strengthen the consolidation"* Alain Brunet et al., "Effect of Post-Retrieval Propranolol on Psychophysiologic Responding during Subsequent Script-Driven Traumatic Imagery in Post-Traumatic Stress Disorder," *Journal of Psychiatric Research* 42 (2008): 503–506.

167 *"A week later"* Ibid., 304.

168 *"gave up his long-time hobby"* Shirley S. Wang, "Can You Alter Your Memory?" Wall Street Journal online, March 15, 2010, http://online.wsj.com/article/SB10001424052748703447104575118021991832154.html.

168 *"the man said he felt remote"* Ibid.

10. War of the Weirds

177 *"hominins show marked gut reduction"* Kim Hill and Magdalena A. Hurtado, "Hunting," *The Oxford Encyclopedia of Economic History* (New York: Oxford University Press, 2003).

178 *"The coincidence of these traits"* Ibid.

180 *"eyes and brains can . . . co-evolve"* Dave Cliff and Geoffrey F. Miller, "Co-evolution of Pursuit and Evasion II: Simulation Methods and Results," in *From Animals to Animals* 4, ed. Pattie Maes et al. (Boston: MIT Press, 1996).

181 *"The further into the upper-right zone"* Ibid., 2.

182 *"The pursuer and evader have the same accelerations"* Ibid.

182 *"In many species of prey"* Ibid.

184 *"once an attack begins"* Austin Sullivan, "How to Survive a Wild Animal Attack," Essortment.com, www.essortment.com/lifestyle/survivewildani_ttsm.htm.

186 *"the survival probability of immobile targets"* Davide Cassi, "Target Annihilation by Diffusing Particles in Inhomogeneous Geometries," *Physical Review E* 80 (2009): 030107-1–030107-3.

188 *"The detailed investigation of these aspects"* Ibid., 030107-3.

11. Attack of the Mutant Zombies!

197 *"blood's lymphocyte cell count"* "Radiation Effects on Humans," Atomicarchive.com, www.atomicarchive.com/Effects/effects15.shtml.

198 *the aftereffects of the disaster* Richard Stone, "The Long Shadow of Chernobyl," *National Geographic* (2006).

198 *"may be more likely to develop"* Mary Shomon, "Chernobyl's Continuing Thyroid Impact," About.com, http://thyroid.about.com/cs/nuclearexposure/a/chernob.htm.

201 *"If there was something"* Stone, "The Long Shadow of Chernobyl."

201 *"I get up in the morning"* Ibid.

202 *"After passing through a checkpoint"* Ibid.

202 *"More than a hundred wolves"* Ibid.

203 *"Animals don't seem to sense"* Stephen Mulvey, "Wildlife Defies Chernobyl Radiation," BBC News, http://news.bbc.co.uk/2/hi/europe/4923342.stm.

203 *"Nothing with two heads"* Ibid.

205 *One study conducted in 2000* Doug Brugge and Rob Goble, "The History of Uranium Mining and the Navajo People," *American Journal of Public Health* 92 (2002): 1410–1419.

205 *"sensitive cells for periods of time"* Ibid., 1412.

205 *"that which gets in the way"* Doug Brugge, Timothy Benally, and Esther Yazzie-Lewis, eds., *The Navajo People and Uranium Mining* (Albuquerque, NM: University of New Mexico Press, 2006).

12. You Gotta Shoot 'Em in the Head!

210 *"Of all the weapons"* Max Brooks, *The Zombie Survival Guide: Complete Protection from the Living Dead* (New York: Random House, 2003), 41.

210 *"Studies have shown that, given the trauma"* Ibid., 50.

212 *"The ultimate in precision shooting"* Robert A. Rinker, *Understanding Firearm Ballistics: Basic to Advanced Ballistics Simplified, Illustrated, and Explained* (Clarksville, IN: Mulberry House Publishing, 2005), 374.

213 *"a trapped woman dispatched"* Brooks, *The Zombie Survival Guide*, 46.

214 *"can damage a barrel"* Rinker, *Understanding Firearm Ballistics*, 72.

214 *"Remember that the special"* Ibid., 244.

214 *"firing both barrels simultaneously"* Ibid., 72.

215 *"The average double shotgun will weigh"* Ibid.

216 *"1/420th of the bullet's velocity"* Ibid, 66.

216 *"An ordinary rifle of .30 caliber"* Ibid., 79.

217 *"Studies have shown that of all wasted"* Brooks, *The Zombie Survival Guide*, 48.

217 *"All else being equal, a handgun"* Rinker, *Understanding Firearm Ballistics*, 318.
217 *"The semi-automatic will usually carry"* Ibid., 325.
218 *"The safety on a semi-auto can work"* Ibid., 326.
219 *"The importance of the bullet"* Ibid., 199.
219 *"there is a partial vacuum"* Ibid., 204.
220 *"As the range increases"* Ibid.
220 *"the spitzer has an advantage"* Ibid., 208.
221 *"often train officers to shoot center mass"* Jonathan Maberry, *Zombie CSU: The Forensics of the Living Dead* (New York: Kensington, 2008), 231.
221 *"A 12-gauge shotgun loaded with"* Ibid., 232.

INDEX